Advance Praise

"I find the book to be quite informative. The author is honest, admitting his mistakes, successes and the lessons he learned along the way. As an investor who rehabs and flips houses for a living, this book laid bare all the pitfalls I have gone through and continue experiencing. The steps the author has listed in the book in various stages of hiring and managing a contractor are clear, methodical and practical. I recommend this book to any struggling investor or a newbie. The book is basically a manual written in a serious but entertaining manner."

—Chris Tuesday

"If I, a contractor, liked this book, then many people in the construction industry should treasure the messages in here. At first, when I saw the book cover "Why Do Contractors Lie?" I said to myself, sure this is another book bashing contractors. Interestingly, I couldn't put the book down once I started reading as many of the examples are real, and the book is written in a very persuasive manner. I have always thought that contractors get a bum rap, but the book gave me a different perspective. It was like looking at myself in a mirror! Until this time, I had never realized that some of the things we do as contractors can be that bad. At least some of us. Now that I have read what people say about us, I have acquired a new sense of purpose and will be more careful how I treat my clients. I recommend this book to other contractors and they will be surprised what they or their friends in the industry do. They will also know what investors and property owners think of them. The information I have read here has forever changed the way I approach my work and my clients. I hope other contractors will also read this book with an open mind."

—Contractor Sam

"I have never read a book that highlighted my mistakes the way this one has done. I am a seasoned real estate investor but I still have a long way to master the contracting process and fully be comfortable working with contractors. Every now and then I think I have learned from a past experience, another contractor pulls a fast one on me! From reading this book, I have realized that it is the "contracting stupid" to paraphrase another political commentator in the 90s. Contractors get away with a lot because property owners do not take the time to learn the tools of the trade in the business they are in. I highly recommend getting a copy of this book if you want a clearly written, informative, engaging, and easy to understand material peppered with real life experiences."

—Raphael T

WHY DO CONTRACTORS LIE?

WHY
DO
CONTRACTORS
LIE?

THE INVESTOR'S GUIDE TO
HIRE THE RIGHT CONTRACTOR FOR SUCCESS

J.O.A. MAURICE

NEW YORK

LONDON • NASHVILLE • MELBOURNE • VANCOUVER

Why Do Contractors Lie?
The INVESTOR'S GUIDE to Hire the Right Contractor for Success

Published in New York, New York, by Morgan James Publishing in partnership with Difference Press. Morgan James is a trademark of Morgan James, LLC. www.MorganJamesPublishing.com

ISBN 9781631950681 paperback
ISBN 9781631950698 eBook
Library of Congress Control Number: 2020933367

Cover Design Concept:
Nakita Duncan

Cover Design by:
Chris Treccani
www.3dogcreative.net

Interior Design by:
Christopher Kirk
www.GFSstudio.com

Editor:
Cory Hott

Book Coaching:
The Author Incubator

Morgan James is a proud partner of Habitat for Humanity Peninsula and Greater Williamsburg. Partners in building since 2006.

Get involved today! Visit
MorganJamesPublishing.com/giving-back

This book is dedicated to my parents:
My mother who worked extremely hard through many difficulties, in my formative years, to ensure there was always food on the table for a family of seven siblings.

My late father, who died in 1984 and instilled in me the virtues of a life revolving around the church and spirituality. The positive lessons he taught me at an early age have always formed the basis of my determination to keep pushing no matter the obstacles. A life grounded on perseverance, honor, courage and optimism.

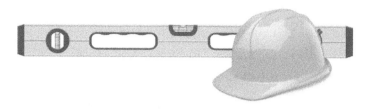

Table of Contents

Chapter 1:

The Pain of Hiring Unsuitable Contractors

Have you ever had an opportunity to be involved in real estate refurbishing in any significant way? Many who have owned homes may have done one form of restoration or another. These are typically one project here, one project there, and may not present major complications. Those in the real estate rehabbing business experience more turbulence than ordinary homeowners can imagine. The moving parts in an actual rehabbing project are many and varied. For beginners, these parts could take a long time to master. Without going through the rigorous process of learning, formally or informally, a new business owner may experience unimaginable stress and loss of investments.

For starters, rehabbing is the process of restoring and improving a property to a satisfactory or superior condition. The process can take weeks to months to years depending on the nature of the work to be performed and the exit strategy desired. The exit strategy could involve renting, flipping, wholesaling, or holding for oneself. Your exit strategy determines how many resources you may put into the house and the amount you can make from the endeavor.

The big picture of rehabbing a property involves purchase, restoration, and renting or selling. The actual steps involved are many and will become clear as you read on. Of these steps, hiring a suitable contractor for your specific job represents one of the most challenging parts of the process. Getting the hiring right can mean the success or failure of your job.

When you talk to seasoned investors, they will tell you that contractors come in all hues and shades. Some nasty. Some pleasant. Some somewhere in between. Whatever shade they come in, don't be fooled by their outward look. Sometimes the ones who look and sound horrible can be the most pleasant to work with. Sometimes the ones who seem friendly and decent can be the nastiest. Please don't let the outward appearance fool you. Look at every individual's total picture, as you will learn later in the book. In other words, evaluate the personality and the contents of the application. After many false starts and deep disappointments in my business,

I finally found my footing. The story of my real estate journey in the next chapter tells it all.

Diagnosing Rehabbers' Main Problem

At a local real estate meeting in Baltimore, a speaker wishing to develop a consensus of what troubled investors the most in the rehabbing process asked questions to over one hundred attendees. Captured below are the answers from one of the investors, John Jacobs, which were in line with the majority of the answers from the other investors:

What drives you crazy in this business?

That many building contractors lie often, and I keep losing my investments because of them.

What worries you the most and keeps you up at night?

The fact that I keep losing money when rehabbing and also the struggles with contractors during construction. I have researched my problems and have come to realize one of my weaknesses is the wording of my contracts with the contractors. My contracts have loopholes which the contractors exploit. Many of the lying contractors know how to twist some of the wordings of the contract to get off the hook when they breach the engagement terms. My concern is that I am not running my business efficiently.

In fact, in one of my current rehabs, I have fired two contractors. The third one who came with a lot of promise now only shows up twice a week. He doesn't even pick up my calls. This contractor calls when he wants to. Now I'm wondering if I am the problem or if it is these contractors?

What are you most frustrated by, John?

That I have been in the business for many years and still make rookie mistakes because I started and continue to run my business the wrong way.

If you were to complain to a friend, what would you say?

I am stuck in my real estate project. I am now on my third contractor in this project, and things are not going well. I am not satisfied. I wish there was someone who could take over the project and continue it to completion. At this point, I don't care how much I pay a consultant. I just want the property finished.

What do you think the cost of not solving this problem is?

Sleepless nights, floundering business, and money in terms of holding costs—to pay taxes, heating, utilities, and general property conservation.

What do you Google when trying to solve this problem?

1. Investor support organizations
2. Suing a contractor
3. Breach of contract
4. Choosing the right general contractor
5. How to find a reliable contractor
6. Hiring a licensed general contractor
7. Contractor red flags

The Main Problem Defined

There were many varied answers, but a consensus seemed to point to a situation that the speaker summarized as, "I need to know how to find good contractors who'll get the job done on time so I don't lose my investment." Again, as this investor has realized on many occasions in his real estate career, the caliber of a contractor can make or break an investor. A pleasant contractor who does not finish a project on time is just as horrible as a nasty contractor who may finish projects on time but ends up stealing materials or tools from an investor. So where is the middle ground, one may ask? Strictly speaking, the answer must be that it is the contractor who gets the job done well and on time. You have to learn to cut off the in-between crap. Be focused as a business owner. The judgment should strictly be on the results and completion time. However, the speaker in this meeting said, many things in real estate are neither that simple nor that straightforward. There are many grey areas that you learn to navigate through with experience.

When you view your real estate journey, picture the path as paved with thorns. Your question and answer should be: how should I walk on thorns? Cautiously and methodically! There are no two ways about it. As the speaker said: verify, verify, and keep verifying both your steps and the people you interact with in this business.

The high dollar amounts involved in real estate investing dictate that an investor proceeds carefully and meticulously. Apart from hiring and managing contractors, you will be dealing with realtors, mortgage brokers, your city, government regulations, attorneys, tenants, and managers. A breakdown in any of the processes or dealings can result in a massive loss of income unless corrected in a timely manner. That is why many people don't recommend starting a real estate business without some formal education in investing in real estate. In the absence of that, they recommend getting a mentor or being involved with some form of consultancy or another. You need to understand that a high dollar business like this may also have a high number of opportunity seekers who may take advantage of you as a newcomer.

At one point, the speaker approached one of the participants and asked if she had tried other ways to solve this problem of finding a good contractor who'll get her jobs done on time so she doesn't lose her investment. "Everything," the participant responded quickly and added that she now feels lost in the business. She said she still gets the hiring wrong, going by the caliber of contractors she gets. At that point, the speaker told her she may need a one-on-one with a seasoned investor, and he suggested they have a discussion at the end of the session.

Still, at the same meeting, the speaker asked everyone who has ever invested or considered investing to outline what their dream come true was. After perusing many answers, he again picked on John's answer. It was in line with the general idea of many in the group: to hire great contractors that get my real estate projects done on time. Many in the group agreed that if someone can get a good contractor, put in place the right policies, and manage the process well, then money will come. Such well thought out and properly designed policies will, in many cases, minimize the pain that may come during the rehab process.

The speaker said the initial thought process in any new business should not be to become profitable from the get-go, but to put in place proper business and management procedures, and then money will come. He also said the weakness of many new rehabbers is to worry about making money straight from their first rehab. This focus on profitability takes away from building your business on sound structures and understanding. These sound structures and understanding become your solid foundational bedrock that contributes to your future peace of mind, longevity, and profitability.

Another desire for many in the room was to learn the importance of managing the investor—contractor relationship in a way that reduces the deep mistrust that sometimes occurs between the contracting parties. To this, the speaker presented case studies of some successful investors. From analyzing the paths of these successful investors, you will realize that many of them spent time upfront learning about the business. "Knowledge is power," people say. The consensus is to spend time learning about the business you are in so you can hire the right contractors. That successful path may take engaging consultants to help navigate the rough waters until you know what you are doing.

Another way is to be involved is continuing education and spending time with like-minded individuals through membership in the local and national chapters of the real estate investment associations, meet-ups, and other networking groups. Attendance at such group meetings may also keep you updated on the regular changes in the real estate environment, including the basic legal requirements of the industry. You may also learn some key characteristics of a good contract, as John had mentioned that this is one of his problems in the business. Contracts are covered extensively later in the book. Poorly structured contracts represent one of the ways contractors will cut corners and lie to you just because they can, and also because they can tell from what you accept of the wordings of the contract that you have little knowledge in the business.

The bottom line is that you may experience problems as you commence your real estate investing business when you start on shaky grounds. When you start on the wrong footing without knowing where you want to start and end, you may experience a myriad of problems that may include sleepless nights, floundering business, and loss of money. However, if you start well, with the right education, consultancy and a good team around you, you will progress well. You will get grounded in the correct methodology of hiring the right contractors who will get projects done well and on time so you don't lose your investment.

Chapter 2

My Journey with Contractors

W hy do many contractors lie? Why do even some decent contractors lie? These are the questions that have, for a long time, haunted me and many of the investors I have interacted with since starting my investment career over fourteen years ago. I am always shocked when I visit investors at their job sites and other venues. The one key overriding complaint is on the challenges of hiring and retaining good contractors. As someone who has been in the real estate investment trenches for many years, I can state with a deep sense of conviction that my business took off once I was able to resolve my questions dealing with hiring and retaining suitable contractors. I can also state emphatically that other investors have gained and will continue to gain from my research as they seek to better handle contractor and contracting issues.

The year 2006 was pivotal for the U.S. economy. The first part of the year recorded a hugely accelerated growth. The second part saw a significant slowdown. Since the U.S. economy had been experiencing tremendous growth for the prior three years, people were spending money as if the economy would grow and grow, ad infinitum! Real estate went through the roof. These were the days when you wondered how people qualified for houses. Many people who wanted to buy a house got one or more.

It was a time of creative financing, creative appraisals, no doc loans, and stated income. In many instances, when you wanted a house, you didn't have to put money down and you didn't have to prove your income. I, too, got into the game and experienced the ups and downs of the market. It was a rough ride, but I rode it with optimism and fear! Optimism that I was in a business line I had always admired. Fear because people were buying houses too fast, and as a keen business student, I suspected that this level of growth was not sustainable. Somewhere along the line, a burst would come. Nevertheless, many people were joining the business, and I knew you could only win if you were in the game. I stupidly stayed on and played along!

I was forty-five years old, with a Master of Science in accounting, and a number of finance courses under my belt from a prominent school within the state university system of New York (SUNY). I also had a bachelor's in business administration from a liberal arts college in the Midwest. I had also taken a number of doctoral classes online in business administration (DBA). In general, I was already well schooled in the field of business and finance before starting on this investment route. Also, before setting up my business as a real estate investor in 2006, I had worked in the corporate world in various capacities for over twelve years—some in managerial positions. My experience in the real estate investment field started when I worked in a real estate investment company (REIT) for six years. I saw first-hand what was going on in the real estate investment world. Also, during three of those years, I helped a friend manage a start-up in a real estate and rehabbing company on a part-time basis. Experience in this small company was an eye-opener, since it gave me the opportunity to interact with contractors, realtors, and other investors.

In the REIT and my friend's company, I rubbed shoulders with teams that comprised of contractors, attorneys, mortgage experts, and other real estate pro-fessionals. In my opinion, I was seeing all the aspects of real estate investing. In 2005, I resigned from my REIT corporation job with the goal of starting my real estate rehab company. I thought I had seen everything I needed to know to start my company. I was wrong. In mid-2006, right before the noticeable beginning of the major economic and real estate downturn, I bought my first investment house. It was a turnkey investment with a tenant and little touch-ups to do. A turnkey investment is a residential property that an investor may buy with tenants already in it, so he starts making money immediately without doing much work, if any. I started collecting rent the following month right after closing on the deal. The turnkey was a good start, but I needed experience with an actual rehab. Soon, I bought my first rehab, and then many more houses within a short period. The going seemed good, or so I thought! It appeared like my REIT experience and part-time work in my friend's business was beginning to pay off. I had believed my experience, especially in the REIT world, would easily translate into a personal thriving real estate investment empire. This was true for a short time.

As a finance student, I quickly learned the concept of using other people's money (OPM) to advance my real estate business. Using a combination of my savings, home equity line of credit, and hard money financing, I bought ten houses in quick succession, within one year. My business was booming. I was on a roll. My experience in the REIT world was paying off big time. My interactions with tenants, contractors, banks, and other stakeholders were spot on. From the look of things, I was on the road to buying many more houses, selling some, renting others, and the circle would repeat. For two years, all aspects of my business were moving smoothly like a well-oiled machine. Money was flowing in. Everyone under my business umbrella was working in unison.

I met and exceeded my construction deadlines. I bought and rehabbed properties, rented, sold, managed, and bought some more. My team was a happy one. A team in a rehabbing or a real estate investing business is a group that an investor can have at his disposal to call on for work or consultation. Such team members may include general contractors and other specialized contractors, like plumbers and electricians, you can call on as needed. You may also have accountants, attorneys, financial advisors, bird-dog people, et cetera.

Then in 2008, at the height of the economic downturn in the U.S., property values started plummeting. All of a sudden, I got stuck with these beautifully constructed houses, unable to sell, as the after-repair value (ARV) was now below the total cost. In real estate language, most of my properties were now underwater. This meant that to sell the properties, I had to get permission from the lenders to allow for sales below the market value—a practice called short-selling.

In many cases, this meant the lenders would not receive all their investment in the properties, and I would mostly lose whatever I had put into the project. I tried to hold onto some completed properties believing the market would soon rise, as in other past housing downturns. I was mistaken. This was not just a housing downturn, but a worldwide economic mess. I thought of renting some of the houses I had rehabbed, but like many real estate professionals know, houses for rent and for sale are rehabbed differently. I was stuck. Houses rehabbed for sale can sometimes have delicate and expensive items, and many tenants are not known to properly care for such things. Homeowners can handle such items with care, as they know they are on the hook for their repair. It is their own home and they are responsible for repairs. I had to let most of my team members go. Many real estate businesses also downsized greatly.

I offloaded many of my investments at below market value, only remaining with those I had paid off or acquired using mainly cash deals. Soon, my portfolio reduced from twenty houses to five. I had put all my resources into real estate. I was devastated. I wanted to quit real estate, but chose to stay because I don't believe in quitting, and I also wanted to use the mistakes I had made as a springboard to make me stronger. Through a combination of painful financial reorganization and re-inventing myself, I managed to have a fresh start.

This time, the goal was to utilize less debt and more equity through partnerships. The challenge at this point was that, with less money and extremely controlled spending, it was not easy to get any form of a meaningful team behind me. Teams require money and a sense of assurance that they are in a thriving business. The hope of longevity and seeing a thriving business keeps team members together and captive. No one can blame them. Who stays put in a sinking ship?

My Trials and Tribulations with Contractors

With my team gone, I had to rely on temporary contractors who I hired for specific projects. These contractors did not owe any loyalty to me or my company, except for the specific jobs they were hired to do. The result was that, as soon as a contractor landed a job with me, they were busy looking for another job, so when they were done with mine, they had another job waiting.

Sometimes, they signed up with more than two jobs, so they showed up at my job one week, another job another week, and yet another one the following week. By the time they got back to my job, they had lost their sense of continuity. I was now operating in a completely unfamiliar territory where I had lost my sense of control. How could I run a business where the contractors I had hired seemed to be in the driver's seat? Sometimes these contractors confused my designs with those in other companies they worked for at the same time as mine. In other cases, they couldn't remember at which company they had left their tools.

I wondered why a contractor would not simply tell the truth that from this date to this date, they would be working at a different site. Their lies led to mistrust, arguments, bad blood, and jobs done poorly. In one instance, a contractor I had worked with for many years persuaded me to advance him over $40,000 that it would make the construction cost cheaper as he would buy materials in bulk to reduce acquisition costs. In economic terms, he would get discounts through bulk buying, and the economy of scale savings would trickle down to me in the form of reduced material costs. He disappeared after he took my money. Mind you that this was a contractor I had worked with in many projects, over a period of six years. I was confused and beyond shocked.

To me, this guy had become more than a contractor. He was now a friend, a confidant at that. I had developed a level of trust with him, trust that many people only have with immediate family members. To say that I was confused and shocked is an understatement! I was mad. I was livid. I felt betrayal beyond measure. I didn't know who else I would trust in this business. I had to use my knowledge from the corporate world to get most of my money back. Still, it was a lesson that you have to be on your guard in this business. You need to have proper rules of your own, and you need to follow the rules even if your dad is the contractor. Business is business! Trusting someone does not mean breaking your rules. Rules are rules. Period!

From all these instances, I thought that there had to be a better and easier way to work with contractors. I also thought that, if even with my advanced education and extensive experience in the real estate business, I could be lied to in this manner, then people with less education and experience were unquestionably at the mercy of these contractors.

I talked to various professionals, attended seminars, read books, and researched various systems that would create a streamlined method to help me with investor and contractor interactions. In the process, I solidified my belief that relationships with contractors are a major problem in the real estate business. In my research, I Googled, "How do I get the support I need to rectify contractor mess?" "Why am I always having problems with contractors?" "I need to find the right contractor," and "I need to learn to detect and quickly fire a lying contractor."

As many experts and investors in the industry will tell you, the success of a rehab or building process greatly depends on your relationship with contractors. Therefore, those new investors and homeowners seeking success in the industry need to know the best ways to hire and retain reputable contractors.

Painful Lessons

The road to riches in real estate can be rough and meandering. There can be a lot of turbulence. You have to be extremely focused and know what you are doing. Any stupid mistake or misstep can be disastrous. As I have learned from my story of many failures and successes, and from other investors, I have realized that if someone stays put on the long and meandering road, they are likely to end up in the bush. Success in real estate requires a clearly defined road map with strategic benchmarks. For example, when designing a rehab project, it is necessary to first understand what you want to do, then decide who will work with you on that project, and the materials to buy. It is also necessary to have an idea of what the finished product should look like. That way, you can be in charge of decision making from the beginning, middle, and end of the rehab.

Sometimes, such knowledge is not easy to gain. In my case, I had to take various classes, like those offered by some national and local real estate organizations. In the past, these organizations taught at a designated place in a city for three days, then asked you to pay for weekly private phone coaching. It was not enough. In many cases, I ventured into various real estate projects without the necessary knowledge. I paid dearly in terms of mistreatment at the hands of contractors, and even from investors who came into my life as mentors, only to sell me their rundown houses and then disappear. I sustained heavy losses in my investments. Frustration after frustration followed.

Wow, Some Contractors Can Be Heartless

My, my. Sometimes you have gone through a lot, you think you have seen it all. As I learned earlier on in rehabbing, no two houses are the same, no matter how similar in structure they may appear. Now, I had also believed that no two contractors are the same, no matter how similar in approach they may appear to be. A friend of mine would say, there is nothing new there. No two human beings are exactly alike even if they are identical twins. Except with contractors and rehabbing, you are dealing with your hard-earned investments, so your emotions are likely to run deep. The lesson here is that every contractor is unique and your only way out is to set ground rules that you strictly go by, save for a few minor deviations depending on different circumstances.

From time to time, I experienced how merciless contractors can be. My first mistake was not knowing the specifics of my final product. I let the contractors tell me what they thought I wanted. Each of the contractors I brought in had a different take on the rehab. They probably realized they could manipulate me due to my ignorance. And manipulate they did! They gave me plans of what they believed was pleasant to hear, but not necessarily what they intended to do. When they showed up to work, they cut corners as much as possible. Since we didn't write out specifics of the project, they left themselves wiggle room to say that what they did was what we had agreed on. They asked for additional money if I wanted to do more to fit what I had wanted from the beginning.

I learned my lesson that it is important to spend time and resources upfront to be able to control your destiny, or unscrupulous people will control you and your resources. When someone is controlling you, sometimes they put their interests first and yours last. In many cases, they put you in a losing situation or one that minimizes your gains as much as possible.

How I Did a Strategic Retreat

Some retreats are permanent and fatal. Others can be temporary and designed with the goal of bouncing back up. In this case, when I took a strategic retreat, I buckled down and started learning from scratch. My first lesson was on how to manage contractor hiring and relationships. The years 2008 to 2010 were tough for me and others in real estate. Most of the properties that people bought in the year 2006 were underwater by the time 2008 rolled around. The U.S. economy was experiencing tough times. With property values down, and the lending rules changing, it was hard to refinance or sell rehabs. I, like many people, got stuck with properties. Painful financial reorganization, including surrendering some properties, stared me right in the eye. I did both just to reduce my stock and re-group to a manageable level.

My new strategy was to buy mostly using cash deals, or in case of a mortgage, give a sizeable down payment so I would have a low monthly payment. Excessive debt was the problem from 2008 to 2010 for many investors. My experience was that an investor with excessive debt generally concentrates on debt management at the expense of sound control of his investments. He slowly descends into a survival mode instead of a thriving path. The results are mostly disastrous.

Using Consultants and Getting More Education

A business professor at one of my universities once said there is no such thing as free lunch. There's always a price to pay somewhere. What matters is whether someone pays upfront or at the backend. But the truth of the matter is, there's always some payment attached to different situations. The good professor was fond of finishing a business management class that way. I learned and confirmed this reality the hard way when I started in real estate. I had a tight budget that did not involve extensive initial educational expenses in real estate. I also didn't have a budget for any form of consultancy. I was there alone, in many cases, with and among the "hyenas," so to speak.

The hyenas were, in many cases, those who presented themselves as people who would help you, only to try to figure out if you had good credit they could use, or if you had enough resources to buy their rehabbed houses or those properties they were putting up for wholesale. Wholesale properties are those that some people purchase at a low cost, then put some markup and sell to you without doing any rehab. This is perfectly legal, except that some investors don't tell you upfront if they want to help you as a newbie or if they just want you as their customer. The proper way is to disclose this conflict of interest before you engage each other in a business relationship.

Real estate investing is a cutthroat game. Whoever you chose as your mentor wants to sell you a house that, in many cases, has some hidden defects or issues like water in the basement. They try to sell you such houses during dry seasons when you may not detect the water problems. In a way, someone has to be aware of the vested interests of a mentor or consultant. Free advisors come, but they want to sell you something. Their services are not free. It is always a good idea to allocate a budget for education and consultancy so you don't go for free stuff with your eyes closed because you have no resources for education or consultancy.

A preacher at a local church in Baltimore, I don't remember who, once said, "Be careful of free stuff. In many cases, free is never free." I was taken to the cleaners for believing that there is such a thing as something being genuinely free. The free mentors sold me their houses at exorbitant prices, and in many cases, the houses needed repairs sooner than they had made me believe. When I confronted them with these problems later, they stopped picking up my calls. Some stopped talking to me.

Soon, I realized the best way is to budget for education and consultants in the business structure. Many established and successful companies budget for Research and Development (R&D) amounts in their annual budgets. My business operation started to change once I came up with an inclusive budget that realistically gave me wiggle room for education and training. I now had a consultant and a team for every question I needed answers to. My investment life became manageable in ways I had never experienced.

Have a Plan and Work that Plan

One of the lessons I learned in my real estate educational journey was the importance of having a real estate team in place before starting a rehab. Some members of this team became my paid consultants. The team should be composed of a real estate attorney(s), realtors, mortgage and private money lenders, accountant(s), financial advisors, and contractors of all kinds like plumbers, electricians, and general contractors. In addition to having a good team, you should be a regular at different real estate association meetings, and you could be a formidable investor. The goal of having a team is to make sure there is readily available expert advice for every question an investor may have without wasting too much time or groping in the dark, so to speak. From my educational classes, the instructors and speakers emphasized the importance of having a plan and working that plan. As Yogi Berra once said, "If you don't know where you are going, you might wind up someplace else."

I learned the importance of starting a real estate business, like any other business, with some basic knowledge and a plan. A plan provides a road map so there is no guesswork. You just need to be disciplined and stick to it. Also, a clearly written and communicated plan can be incorporated into a contract that guides the conduct of all the contracting parties. A contract may also provide for inspection and review points that can unearth mistakes in a rehab in a timely manner to be rectified.

As you will note in this chapter, I have had a long experience starting with venturing in this business without the necessary required knowledge, making several mistakes, then stepping aside, taking a much-needed strategic retreat, then re-starting my career in the right way. This re-starting involved factoring into my budget amounts and time for education so I could build my business on a strong foundation, the way you need to when starting. If you are already in the business and experiencing turbulence, you may need to take a strategic retreat and re-group. I hope my story inspires you to re-start in the right way like I did.

Chapter 3:

From an Insecure to a Confident Investor

D o you ever wonder why some people feel insecure in the daily operations of their businesses while others seem to manage just fine? Have you ever stopped to think why some people lose money, lots of it, year in, year out, while others seem to move along methodically, with less struggle?

As a newbie or someone who has been in any business and still struggling, these are questions you may confront from time to time. In real estate investing, since people deal with big dollar items, business challenges can be daunting. There are usually many moving parts to real estate investing. There is buying, rehabbing, renting, and selling. In all these processes, you can say that a contractor plays a pivotal role at every stage. A contractor could help with analysis before buying. A contractor could be key to your success or failure in the rehabbing process. A contractor is key to your maintenance issues in the rentals. A contractor could also help with the building design that keeps prices low so you can make a profit when you sell.

In the field of real estate investing, mastering the art and science of dealing with contractors can make or break your business. Many contractors are aware of their enormous power in real estate, and they wield it to maximum advantage no matter your situation. Some contractors are decent and they will advise and work well with you. Some are horrible, to say the least. Many are somewhere in between. The challenge is determining which contractor you end up working with. That's the six-million-dollar question. That's the question this book will help you answer.

This chapter outlines the framework that you will need to go through to learn to work with contractors in amicable and less confrontational ways. In all facets of life, starting a new business can be scary, even to a seasoned businessperson. Like in any business, some education is necessary, whether that is through a formal classroom structure or through the school of hard knocks. Street smarts, as some may say! In a formal classroom setting, or even through other media outlets like the internet, you usually pay upfront unless you get some funding or waiver. The key thing in this setting is you are paying upfront unlike the school of hard knocks, where you leave yourself exposed to the world to teach you. As many street-smart learners will tell you, the world can be unforgiving. If you leave yourself at the mercy of the world, you may have to go through some tough lessons that may take a long time to learn and may involve heavy, heavy losses before you find your footing.

This is what some people call paying at the backend. You lose first, then double pay to cover your losses and then some. When paying at the backend, getting in may not be that expensive, as you cut out initial educational expenses, but you will pay for your mistakes dearly. Typically, the choice is with the investor. This is how I got into real estate, and I paid dearly for setting myself loose to the world to teach me. I went into business without the necessary education first. To this day, if a new investor comes to me for advice on how to start any business, I usually stress the importance of including the educational and consultancy amounts as part of the start-up capital. This way, they prepare themselves to hit the ground running, armed with some knowledge that will help them make informed decisions from the get-go when dealing with sellers, mortgage bankers, and contractors.

Getting some education before starting a business and getting familiar with the landscape in which you and your business will operate is key to answering these basic questions: Do you ever wonder why some people feel insecure in the daily operations of their businesses while others seem to manage just fine? Have you ever stopped to think why some people lose money, lots of it, year in, year out, while others seem to move along methodically, with less struggle?

Relevant business education and familiarity with the landscape in which you will be operating provides good grounding. As opposed to people who may start their business without the necessary knowledge, you will have your foot anchored in solid foundation and not in sand.

From my experience and from talking to many real estate investors whose businesses are struggling, there is usually one overriding problem they struggle to solve: they need to know how to find good contractors who'll get their jobs done on time so they don't lose their investments. The process of interviewing contractors so an investor ends up with good ones is a major dilemma for many investors. The majority of the investors have one simple dream: to have great contractors that get their real estate projects done on time.

Although the field of real estate investments is large and varied, the area of rehabbing may be narrowed down to three key aspects: contractor, contractor, and contractor. Indeed, many rehabbers dread the thought of dealing with contractors, yet many people who cannot build properties on their own have to use them. It's kind of a love-hate relationship. You hate the contractors, but you need them.

As you look back on any building projects you or anybody close to you have undertaken, you may want to see if you have had the same questions on how to find a good contractor and if you shared the same dream outlined above. This book will help jog your memory to see if you have had the same questions and how you handled them. These questions should be helpful to you:

- Do you dream of hiring great contractors that get your real estate projects done well and on time, but have no idea where to start?
- Do you constantly lose your investments because of hiring contractors who are not suitable for your jobs?
- Do you know that it is possible to learn a streamlined, practical and well thought out methodology for finding the appropriate contractor for your jobs?

To help you answer these questions, this book will give you a framework of how to approach the contracting process with ease and confidence.

This chapter presents the road map of how you proceed to master the process of hiring the right contractor, who will finish your projects in a timely manner, so you can stay in the investing business. The steps below, when followed sequentially, should place you in the right path of profitability with peace of mind:

- Find a contractor you can trust
- Spot when contractors start to lie
- Make sure your project gets completed perfectly on time
- Avoid end-of-contract arguments or lawsuits

Finding a contractor you can trust starts with knowing what you want the end product to look like. You also need to know the landscape in which you are operating. For example, what do the kitchens in the newly built houses look like? How much does it generally cost to build or rehab a house like the one you want to purchase or already purchased? What types of materials will you need, where do you buy them, and what might they cost? Other questions include which type of contractors you need for this project and how you find them. These questions, together with some decent skills of hiring contractors and concentrating on their track records, can help you zero in on the right ones you can trust from a crowd of many.

Spotting when a contractor starts to lie takes some experience and skill. To even have a shot at this skill, you need to ask yourself if you are good at spotting liars in general. A discussion with a psychology expert might reveal some tips. Such tips include lack of direct eye contact, playing with the fingers while talking to you, constantly repeating some phrases and biting the lip repeatedly. Now, you don't have to be a psychology expert to figure out contractors who lie

While some of the psychology methods of detecting liars may work for you in this context, there are some specific real estate pointers: trying to rush to consummate a contract and always talking about the pictures of their supposedly past work in their phones when you ask them a question. The pictures of past work on the phone or a camera can be good to help gauge someone's capabilities, except that you may not know if that was their work or if they just took pictures of someone else's work. You need to ask additional questions to verify the authenticity of the photos. For example, you could ask how long the work took, how many people worked on the project, if he still works with the investor—if not, why not? You may also ask for the project's total cost. If the contractor appears edgy and evasive, then you may not rely on the pictures for your decisions.

You could also spot when a contractor starts to lie when after signing a contract and taking the down payment, he comes up with several amendments that he claims he just now realizes should be amended in the contract or stricken out. He misses work, without real reason, on the day you propose to have a major meeting to sort out work issues. When you ask him a question, he blames his employees or subcontractors when he's the boss. When you meet him at the project site, he concentrates on issues unrelated to the work at hand. These and some other indicators you will discover as you run your business will help you spot liars before they mess up your business too deeply.

Sometimes, the lies may be serious enough to contravene the contract. Sometimes, these lies may raise a red flag signaling that you need to be careful when dealing with this contractor. Sometimes the lies are too many, it may be a clue to hire another level of supervision so there's more than one person dealing with him. Many liars fear a crowd because more than one person is a witness to their lies. Sometimes, you may need to tape your meetings with him. Sometimes, you may need to document construction progress in stages so someone is not pointing to what they did last week as having been done today. At any rate, once you spot the lies, you can decide how to handle them according to the prevailing circumstances, which you will be aware of as you continue reading this book. Some lies, like those involving massive theft of your resources, can be documented and may be too damaging that you need to terminate the contract instantly.

After hiring a contractor, or even before the hiring, you should have an idea, a road map so to speak, of how the project will begin and end perfectly on time. This expertise could come from training, experience, or consultancy. Assuming you have done your homework on how to find a contractor you can trust, proper work progress requires making sure the contract is watertight and there are specific inspection and payment triggers and times. You also need to outline contract termination procedures and conflict resolution mechanisms. When all these procedures are in place, the web that combines and allows for smooth interaction is communication. Make sure you clearly state how communication is to happen: how many times you may need to have meetings, inspection times, payment schedules, how to amend the contract, and chain of command if working with different layers of contractors and subcontractors. Also, you need to make sure the contract includes clear communication procedures and properly defined timelines. Always, as the investor, you need to continually pay attention to the task at hand and not take your eyes off the ball at any time. This close supervision can ensure quality and timeliness, and results in a properly completed project.

A project that is well-done according to the contract and ends on time minimizes end-of-contract arguments and potential lawsuits. All the moving parts covered so far are interrelated. When you hire well, you minimize lies because you weed out the liars from the get-go, and even if they sneak through, you are able to spot them and handle them promptly and accordingly. Clearly defined have a streamlined, practical, and well-thought-out methodology for finding a contractor with a good track record by asking the right questions for different types of jobs, seeking and verifying appropriate documentation, getting detailed written estimates, and managing relationships with contractors properly. This book provides a reference point for all those wishing to enter into contractual obligations with contractors so they have the necessary knowledge to approach the contracting process with ease and confidence.

Chapter 4:

Master the Real Estate Landscape Around You

Boy, I love weekends. I cannot wait! The cookies, the coffees, the balloons hanging on the mailboxes. The smell of new houses. The different designs. The imperfections. The couples. The empty houses. It's all fun!

My eleven- and nine-year-olds have also become interested in the business and now we move around on weekends as a team. Who knew I would co-op these kids to be my buddies on weekends when their friends are hooked on the electronic gadgets?

Sometimes, they will go to their Saturday morning games, then we would grab a quick lunch at home, then head out. Sundays are mostly the same. Church in the morning, open houses in the afternoons.

For beginners, these are called open houses. From the spring, through the summer and fall, you are likely to see these sights in your neighborhoods. People buy and sell houses all the time. For those in the market to buy or sell properties, these are treasure troves of information. If you want knowledge on building styles, appliances, plumbing and electricals, fencing and landscaping, painting and the types of windows and doors, then you may want to frequent these staged houses, as some people call them.

You also get a chance to interact with real estate professionals like realtors, mortgage brokers, investors, and maybe contractors in a few cases. If you are a builder, you may ask these real estate professionals for contractor contacts.

In John's situation, he made these visits a family affair. This was one way to interest his family in the rehab business. They would visit and discuss different projects together. Visiting these houses also helped him understand certain scenarios when discussing his plans with contractors. He even got ideas of where to buy materials cheaper. To this date, John is hooked on visiting open houses at least twice a month.

However, visiting open houses is not always fun. In his lifetime, he has visited some awful houses. He wonders how some people don't care what they put out there. He thinks these are either ignorant people with some lousy realtors, or they are those who are disgusted with their houses and just want to move on. Even in these situations, he learns how not to stage a house for sale. He tells his kids all lessons are good. A poorly staged house can be a great bargain. The condition of a house can determine how much people are willing to pay for it. He believes that, as a seller, you get what you put into it, whether that is money or time! John learned to view each house condition as a lesson and an opportunity in itself.

Despite some of his disappointments, he enjoys socializing with the professionals he meets at the open houses. He has learned a lot from them, made many contacts, bought and sold houses through them. He is hooked! This is part of his research. He has now dubbed the open house tours as opportunities for research with fun.

Given that buying or renovating a house is a key investment in people's lives, many training institutions and seasoned investors stress the importance of devoting time and resources upfront for research so the building process starts well, flows smoothly, and ends as planned. Research could take many forms such as open houses, reading various publications, a casual conversation with contractors, googling different plans, knowledge of the local property prices, and even coming up with a sketch.

Extensive research—as much as possible—is necessary before deciding and proceeding with a renovation project. Research may lead to ascertaining the space to build on, design features, contractor hiring, budgeting, for various reasons. Many are home buyers, some are window shoppers scouting for the latest designs, some go to pick the brains of realtors, others to seek contractor contacts, and a few others mainly to talk to those who are looking to establish what the local community is up to. Visiting open houses presents you with an opportunity to see completed projects. You will obtain an informed and realistic knowledge of various possibilities or angles your project may take.

Interactions with Realtors

As an investor doing research on the local real estate environment, talking to many realtors at open houses provides an avenue for a free-flow of ideas in a relaxed atmosphere, as realtors are known to talk freely here. In these settings, they are not just looking for immediate buyers, but also referrals and future prospects. Other investors approach realtors to see what's in the market for sale, get knowledge about where the market may be headed, and also to get an idea of the comparable properties.

Latest Designs

Others visit open houses to look at the latest designs of kitchens and countertops, closets, bathrooms and fixtures, and even current painting colors. Real estate agents may also suggest other looks that they see in other regions that may work for an investor in their region.

Contractor or Designer Referrals

Realtors do have contractor and designer contacts that they interact with. So when doing research locally, you could ask a realtor to put you in touch with the contractor who has designed or built the structure you are viewing. Many realtors will be glad to provide contractor names, as the house you intend to work on may be the next one they sell.

Google the Different Building Plans Available Out There

Through Google or other search engines, you can develop a basic idea of the type of architectural or design plans that interests you. So when you approach the professionals, you are not a completely blank page. Many people tend to treat others with respect when they believe you are knowledgeable about what you are discussing:

1. Floor plans: you get to see everything in your space, like how the furniture fits. You get to look at a plan as if the ceiling was taken off and you are glancing at the floor from above

2. Site plan: shows the entire building with a plan of the site

3. Ceiling plans: may include architectural plan details such as vaulted ceilings, lighting fixtures, wiring, and even small details as switch locations.

4. Millwork drawings: give precision drawings with exact measurements of designs and fixtures

5. Interior elevation: this can create a view of the interior with baseboards, windows, crown molding, kitchen or bathrooms.

6. Exterior elevations: shows you the outside of the building and what the structure will look like when finished.

7. Landscape plans: shows you what the landscaping space may look like with foliage, small gardens, walkways, flowers, and lawn decorations.

Come Up with a Sketch or Layout

In many cases, it can help if you have a rough sketch of some of the structures you want to work on. You don't need to put measurements in the sketch, just include how you want structures arranged and situated. For example, arranging a kitchen with an island, cabinets, and a breakfast bar may be different structurally depending on the space you are working with and the neighborhood. You can do a sketch so when you are talking to a contractor, you have everything in front of you. Sketches you draw may form part of the contract such that there is no mistaking what both parties are agreeing to.

Become Knowledgeable in the Property Prices Around You

As an investor and a builder, it's important to have some basic knowledge of property values and how to estimate repair costs. Real estate prices can be established by doing a real estate analysis, also called comparative market analysis (CMA). CMA is a study of investment properties in a market to determine its value.

For a rental income property, the fair market rent can be calculated using current market worth. Fair market rent represents the estimated rental amount, which includes base plus essential utilities that a property may rent for in a given area. U.S. Department of Housing and Urban Development (HUD) publishes the figures annually to guide Section 8 contracts and other Housing Choice Voucher programs. Private landlords may also rely on these figures or on percentages applied to the CMA calculated amounts. For example, using a property's value, you can charge rent based on percentages ranging from 0.8% to 1.1%. Therefore, if a home value is on the cheaper side, say less than $350,000, use percentages closer to 1.1%. From $350,000 and above, you may use percentages around 0.8%. Also, when buying and selling, CMA may be used to set the listing prices.

How Brokers, or Other Professionals, May Calculate CMA

They may start by compiling values of similar properties that recently sold in the same locality and subdivision, typically within the last six months. Many factors and physical features are used to select properties to compare (comps). Then adjustments are applied as necessary, as no two properties are exactly alike. The adjustments ensure that the comps being considered are as similar as possible. For example, when the subject property is a two-bedroom home, and the comp is a three-bedroom house, then adjustments have to be made as outlined in the adjustment section below.

Physical Features

These include kitchen, bathrooms, number of bedrooms, central air, and heating system. You also have to compare the same types of properties. For example, are you dealing with a townhouse, condo, or a single-family home? Whichever the situation, make sure you are comparing similar structures and features.

Location

Location, location, location is a popular mantra in real estate investing. Some investors say location is a first among equals when considering the many factors to take into account when buying or renovating an existing property. Location can help with establishing the price to pay on property purchase, rental values, and the type of amenities to include when upgrading or rehabbing. When analysts pick comps, they ensure, as much as possible, that they are looking at properties within the same subdivision and similar areas to the subject home they are appraising.

If unable to find enough properties in the same subdivision and similar area, then they may widen the search to a broader locality and larger area. It's important to note that a broader locality and a larger area may not provide as close a number as properties in the same subdivision and similar areas.

Data

Include items like vacancy rates, the amount for rent, and the price per square footage. Analysts will look at vacancy rates in a neighborhood to gauge the impact on the subject property's bottom line in terms of rental income. The vacancy rate may also highlight the concept of supply and demand. More supply than demand in an area may also mean lower rents. Analysts will also compare rent charged among similar properties to get the base going rent in the area. That way, if an investor decides to add amenities that may end up valuing the property above the comps, then that will be his choice, but he will be aware of the baseline rent he may get. Another factor to consider in CMA analysis is the price per square footage. This is important data for comparison, as it is mostly the same among the properties, ensuring a comparison of oranges to oranges.

Applying Adjustments

These may include, for example, the number of bedrooms and bathrooms in a home versus in another home being compared. If you want to use a property with a two-bedroom as a comp against one with a three-bedroom house, you may start with calculating the price per bedroom, then adding that to the value of the two-bedroom house so you are comparing oranges to oranges.

Adjustments are done to ensure the comparisons are as close to each other in structures as possible. You do not want to compare a three-bedroom house to a two-bedroom house without making price adjustments to equalize the home values as much as possible.

The importance of researching your neighborhood and becoming familiar with overall real estate trends in your area before purchasing or starting a building project cannot be overstated. Preliminary research will provide you with the necessary background for decision making so you don't approach the rehab process from a point of ignorance, guesswork, and rolling the dice with your resources. With this knowledge, you will acquire clarity on what you want to do and the questions to ask professionals like realtors, mortgage brokers, contractors, and even government agencies such as the city and taxing authorities. Also, your knowledge of local styles and structures currently in force will help you become familiar with the building project you intend to pursue.

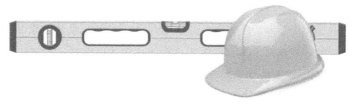

Chapter 5:

Know Your Project Intimately

t's 10:40 a.m. on a clear, crisp Monday morning. People are pouring into this house on the 2600 block of Federal Street in Baltimore, Maryland.

"Well, well, well. Another day, another auction. This may be my day. The day I may finally gather courage and do what I need to do. I have already walked around the house and I like what I see," Ron said to himself as he tried to attract the attention of another gentleman standing nearby.

The gentleman was so busy looking through his phone, he didn't seem to hear what Ron had said. "Or maybe he just ignored me?" Ron said in a low tone as he moved closer to a lady standing nearby.

Ron is a short, stocky looking type person. He is shy and has many insecurities. Talking to strangers is one of his challenges. He gets discouraged easily.

On this day, he was determined to strike a conversation with a stranger. Ladies tend to be sympathetic or at least may not ignore people like men do, Ron thought as he positioned himself near the older looking lady. As a shy personality, he wanted to play it safe. He didn't want to approach a young lady for fear of rejection in case the lady thought he was about to hit on her.

"Are you also here for the auction?" he asked as he looked shyly at the lady.

"Yes. You? My name is Rebecca, by the way," she said as she stretched out her hand to him to shake.

"I am Ron. May I tag along with you so I can learn how auctions work? I am new in the business."

"I'm not sure I can teach you much, since I am also new in the game, but we can definitely learn together."

"Thank you." He was disappointed, sort of, but at least embraced the idea of working together with someone. Ron tried to keep the conversation going so the lady would not walk away from him. "All these vacant properties in Baltimore, and I can't make up my freaking mind. I hope this will be my day. I have been attending many auctions. Some online, others in person."

"I know exactly what you mean. I'm in the same predicament," Rebecca replied as she nudged him to move closer to the front, as the auction would be starting in five minutes at 11:00 a.m. "I hope you have registered for this auction!"

"Definitely." Ron said. "That's the first thing I did when I got here. At least that much, I know. I also walked through the house, so I have a pretty good idea of what I will be bidding for."

At 11:00 a.m. sharp, the auction started. The announcer read the ground rules, responsibilities, and obligations for those winning the auction. He also covered a few other housekeeping procedures, then started the auction from the advertised base: $5,000, $7,000, $7,500, $10,000, $11,000, $12,000, $14,000.

Ron and Rebecca stood in silence. None of them even attempted to raise their hands or shout a number. They periodically glanced at each other, but still didn't say a word. The auction amount was now at $15,000. Going once, going twice, going thrice.

$16,000. Going once, going twice, going three times. Deal!

The auctioneer said as he pointed to the winner. The gavel came down. The auction was over. Others started to walk away as the winner moved to the front to talk to the auctioneers.

Ron and Rebecca moved to a corner to talk. "It's been six months, and I still can't get myself to pull the trigger. I am so fearful, yet there is so much gold in this city. All these abandoned and boarded up houses are gold, you know." Ron said as he gave a shy laugh.

"The city of Baltimore has many abandoned houses. Someone can make real dough from these." Rebecca said. "The city doesn't seem to want to let go of them. They own some, and they are in the process of owning others with so many code violations on them. So much crime in Baltimore due to these empty houses. The criminals have convenient hiding places, I think."

"Not really," Ron answered. "I am confused. A speaker at one of the real estate meetings discouraged the audience from purchasing such houses. He said this is like a war zone in real estate terminology. You need to be extremely careful about what you buy." He was a veteran investor of over twenty years who seemed to speak from a position of deep knowledge. He said he wanted to help new investors in Baltimore in any way possible. Rebecca listened attentively as Ron continued, "Many landlords struggle with renters in these neighborhoods. Houses here appear cheap, but the headache they come with is not even worth it.

"Many tenants here do not like to pay rent. They have that dependent mentality because they are constantly in one form of government assistance or another. They mess your house up, they have no sense of responsibility. That's why I'm scared to buy here."

"Hmm!" Rebecca mumbled as she looked at Ron directly saying, "Now that's confusing."

"I fully agree," Ron replied and asked Rebecca, "So, how many houses have you bought?"

"Three! I don't fool around. I do my research and if the numbers work, I consult with some contractors and buy. Real estate investing is a tough business. Auctions go quick. You can get some fantastic deals, but you have to know what you are doing." Rebecca said. "I have to run, but I will tell you one or two things. You have to shake off that craziness of analysis paralysis. Listening to so many people will keep confusing you. I went through that for a year and finally teamed up with another like-minded investor. We think alike and talk often. We are not partners, but we compare notes all the time."

"All these people you saw at this auction are good, but there are crooks among them too. You have to study people and know who you can work with as a friend. Sometimes you may pay something small for consultation, and that is good. Just list that as a business cost in your accounting and tax books. Be careful. Real estate is a high stakes game. There are *many* good people in the business, but there are also many crooks too. Experience or mentoring can help you sift the chaff from the wheat."

"I was lucky to meet this friend I compare notes with. You don't always meet like-minded friends like that."

"I understand. You have been helpful." Ron said as he thanked Rebecca for her time and insight. "Do you mind exchanging numbers? We can always connect and talk some more."

"Sure, here is my business card. Thank you," Rebecca said as she handed Ron her card and walked away. Ron stood still. He was mad at himself. He wondered what his next step was going to be. This auction was like his last stand. This is where he was to buy his first house. Now the auction had come and gone, and he hadn't even mustered the courage to bid. He realized he needed to do more networking.

That evening, he went to another meet-up. The theme was: you have bought your house, now what? The keynote speaker seemed like a serious no-nonsense businessman. He read mostly from his notes which are summarized below:

If you have ever rehabbed a house from beginning to end, you have no doubt realized that it can be complicated, as there are usually many parts to the whole. As such, an in-depth knowledge is necessary to have a good shot at successful completion without much stress. In rehabbing, investors typically have three forms of exit strategies when complete. They can sell, hold, rent, or refi to lower rates than to rent or live in. The type of exit strategy desired and the market conditions may determine the level of rehab to be performed. This level of rehab is one of the reasons an investor needs to develop an intimate knowledge of his project to be able to hire the appropriate contractors for the appropriate jobs for the different markets.

Looked at in totality, a rehabbing project can be daunting. Some houses that may need complete rehab can be scary and breathtaking. There is the demolition and the trash phase even before the building begins. However, with experience, you can do well in some run-down, scary houses. For newbies, you are advised to start your real estate journey with a mentor or consultant. You can learn a lot within a short time. You may also be able to learn to negotiate with sellers and how to talk to different financiers to fund your different projects. He mentioned that many investors approach rehabbing differently. No one way is perfect. For him, he looks at a property in terms of individual projects that fit together to form a finished house.

These individual projects are wiring and electricals, plumbing, roofing, basement, kitchen, bathrooms, and living and dining room, as many people are into the open space concept. Then he also looks at painting as one project and outside the property as another separate portion. Dividing a house like this helps him with analysis and hiring. He told participants that they should approach rehabbing a property as someone who has to eat an elephant. It is big and scary to look at, and so you have to approach it one limb at a time. Approach rehabbing a house the same way, especially when new to the business. It's big, it's scary to look at, and you have to compartmentalize the house as consisting of different and separate sections in your mind. Otherwise the thought process of rehabbing a whole house can look overwhelming. A good and experienced mentor or contractor can help you go through the emotions.

For those investors who have been in the business for a long time, they will attest that, if someone bought a property at the right price and hired suitable contractors, then the chances of making a killing were mostly high. Rehabbing is where the money is, it's lucrative, but you have to buy and build right. However, if not done right, it can bankrupt someone because of the large investment amounts involved. Therefore, before starting any rehab project, please be advised to assess your finances, experience, and understand the physical environment in which you operate.

Rehabbing can cost anywhere from say $10,000 to $100,000 and more. Some of the common basic costs include permits, labor, and materials. The first step in rehabbing is to assess your capital. This will guide you in terms of the type of property to purchase and the scope of work you may need to undertake. The next step is to tour the locality in which you want to buy the property.

Look for houses that meet your price range in terms of the work that needs to be done. After purchasing the property, or even before purchasing if you are allowed to spend ample time inside, take a good look so you have a rough initial estimate of the budget involved. That way, you may have an idea of whether you even have enough resources to undertake rehabbing now or later. That may also mean commencing a contractor search immediately or waiting until later.

Other factors to consider before purchasing a rehab include assessing how quickly properties take to sell in the market. You also need to evaluate how long you may take to rehab the property. This takes into account the size of the subject property, the individual renovation projects to undertake, and the team of contractors you will be working with. Each of these factors and the others above will give you a good idea of how quickly you can make your money work for you, or if you will be able to repay any short-term loans you may use. You will also have an idea of how much time to request for financing. You need to be aware that short term rehab loans can come with high-interest rates and stiff penalties, and may include foreclosing on the property if you breach the terms.

In general, the process of rehabbing includes purchasing a property, renovating, and either renting or selling at full market price. Rehabbing can be complicated and time-consuming, even to a seasoned investor. This means that newbies may take a long time to master the rehabbing process, and so they should pay super attention to the details of all the moving parts. As already described above, these moving parts include first researching the local market area, evaluating capital availability, and determining who you will work with. Many new investors start from attending meetings in the local real estate associations, meet-ups, and other educational outlets. These groups provide great networking opportunities and a new investor may even hook up with a reputable mentor. Other new investors hire a consultant from the get-go so they start their investment journey with a strong foundation.

In most cases, rehabbing a house for rent is different from rehabbing one for sale. The below outline will help you split the rehabbing process into manageable chunks.

Initial Walk-Through

You can do this yourself if you have the knowledge. Otherwise, do a walk-through with a home inspector or a potential contractor likely to work on the project. You can also pay another contractor to help you out at this stage to create a rehabbing plan. During this walk-through, remember to carry a camera, flashlight, measuring tape, drill, hammer, crowbar, and a notebook. You should get a detailed assessment of the property and take pictures of all sections—concentrating more on problematic areas. Get accurate measurements so you know how to organize. For example, when considering kitchen and bathroom areas, take measurements so you buy appropriate appliances and fixtures. Accurate measurements and detailed photos will help with careful review and the creation of a plan later. The photos will also provide the before and after repair shots for your records. The photos can be helpful evidence in case disputes arise. Some properties have no electric power, and so a flashlight can be handy. Other properties are boarded up and you have to un-board to access and re-board when done.

At the point of creating a plan, you will evaluate the size of the windows you need, whether you want to retain the old ones and their sizes or if you want bigger ones. Evaluate the type of flooring you need, if that is carpet or laminate. For overall improvements, see if your goal is to increase the value for resale or refinancing. If your desire is to rent out, then the rehabbing procedure may be different. You may opt for good touch-ups and not worry too much about outside appeal and other exotic decorations. In all cases, your finances, desire, and neighborhood appeal will dictate how you proceed.

The next stages of the planning process may be crucial, so ask someone with experience to help you out with ideas for property sketch, itemization of proposed repairs, and improvements to the last measurements and detail. This way, the contractor will clearly see your vision when working on his proposals. Such detail will also prevent ambiguities when work starts.

Scope of Work

If your initial walk-through is thorough and the pre-rehab planning process above goes well, then the scope of work phase should be completed relatively smoothly. The scope of work stage is the place to outline in detail the extent of the rehab you need. You will include all the major and minor renovation details. This is where you detail the extent of the project for your contractors to assess the level of complexity you require. Again, this scope is an important planning phase that you are best advised to seek help with if you don't have experience. This is also an important phase as you interview contractors, as the scope of work is what they will bid on. The following steps should be helpful:

Review Your Planning Notes

List all the repairs you want—like what areas to demo, trash removal, installations, sections to replace, roofing requirements, windows, floors, doors, et cetera.

Work on Your Budget

Now that you have a good idea of the scope of work, work on your budget, drawing out the cost of each project. In many cases, you already have your finances in order, so you are basically allocating costs to each individual project. The numbers in your budget should guide which renovations to prioritize, which ones to make optional, and which ones you may need to do away with.

Also remember to include holding costs, as there will be periods between completion and rent, sale, and marketing. Some people make this allowance anywhere between ten to twenty percent of the total rehab cost. This allowance may also cover unforeseen delays and unanticipated problems. Exercise discretion. In your budget, exhaustively detail everything you want done. Include furniture, fixtures, faucets, window blinds, color paint, and many other details, no matter how small they may appear.

Review the materials needed and their cost. See which materials you may re-use and what you need to buy. Call home improvement companies like Home Depot and Lowes for pricing details. Check other stores like the Restore or other thrift stores that may sell used materials at huge discounts. Second Chance and Loading Dock, if rehabbing in Baltimore, are examples.

A properly prepared scope of work and budget estimate will make your negotiations with contractors exhaustive and more effective, as everyone will be on the same page.

To Rent or To Sell

Rentals are rehabbed differently from flips or high-end sales. Some aspects that may determine whether to rent or sell include taxes, rental license fees, and inspections. These aspects may reduce rental cash flow. In this situation, an investor may decide that the best route is to rehab for sale.

Rehabbing cost is another determinant for selling or renting. Rehabbing for sale is typically higher than rehabbing for rent. With a rental, sometimes you are just looking for minor upgrades for an immediate move-in. When fixing to sell, there is usually a need to upgrade the features to attract buyers so they may pay a higher price. When rehabbing to sell, you may choose to concentrate on certain sections more than others. You mostly want to make the kitchen and bathroom more spectacular with good upgrades, fashionable paint, electrical fixtures, flooring, and furnishings. The basement and garage may also improve property appeal for sale.

In all cases, whether rehabbing for sale or for rent, you should check the neighborhood to see the current styles in force, so you are not going way overboard in amounts you may not recoup due to comp limitations.

When fixing to sell, you have to pay attention to the taxes. If you buy, rehab, or sell within one year, you may pay the short-term capital gains tax. After a year and one day, your taxes would be approximately reduced in half, since the profit on the sale would now be taxed as long-term capital gain. Your determination to fix for rent or sale may be determined by the time of completion. If you will be finishing close to the wintertime, the house may be in the market for a little longer until the beginning of spring. You will need to schedule times for rehab for sale differently from the times for rehab for rent. If you have to keep a house during the winter to sell in the spring, then be aware of holding costs that may include heating or winterizing and snow removal.

Rehabbing for rent has its challenges, also. You need to be aware of the obstacles you may face when trying to refinance. Sometimes, when investors rehab for rent, they may cut corners and not pull all the required permits. When refinancing, banks may require verification that all the building codes have been complied with and there is a reliable tenant with at least a year-long lease.

Then there are other rental costs, like management and repairs. Make sure you take all these factors into account when rehabbing for rent. Some banks may only give higher interest rates to rental properties.

Also, you should plan for various exit strategies in case your preferred method doesn't work. For example, if you rehab to sell and you don't succeed, what's your next strategy that will not leave you in a worst-case scenario? A strategy that some investors may employ could include utilizing the sale-lease option after a year and one day from the date of purchase to save on taxes. A sale-lease option releases the investor from management and repair commitments.

Know Market Rents and Comps

The market rents and comps can be calculated as outlined in Chapter 4. For a rental income property, the fair market rent can be calculated using current market worth. See the calculation details outlined in chapter four.

For comps, you may start by compiling values of similar properties that recently sold in the same locality and subdivision, typically within the last six months. Many factors and physical features are used to select comparison properties (comps). Then adjustments are applied as necessary, as no two properties are exactly alike. The adjustments ensure that the comps being considered are as close to being as similar as possible. For example, when the subject property is a two-bedroom home and the comp is a three-bedroom house, then adjustments have to be made as outlined in the adjustment section below.

Also, you may contact real estate professionals like a realtor for market rents and comps. They may have them readily available from the multiple listing compilations (MLS). You may also visit sites like www.rentometer. com, Zillow Rental Manager, Trulia.com, and Realtor. com for local rental comparison analysis.

Determine If You Want to Do a High or Low-End Rehab

That depends on the neighborhood and your budget, and also determines the type of contractor to hire. A high-end rehab requires a different thought process. The clientele will be different and most likely more demanding. The fixtures and appliances will be more expensive and classier. The exterior of the house will be more appealing. The same goes for the landscape too. Therefore, you will need more capital and more experience to do a high-end rehab. The rental income should be higher, and you may get more reliable and stable tenants. Do your research on similar property values in similar neighborhoods, and when you are convinced that you have the necessary resources for purchase and rehab, then go through the process of selecting suitable contractors, as already explained elsewhere.

If in the Rehab Business, Make Sure You Have a Team

As for many successful businesses, even if you are a sole proprietor or part of a partnership or corporation, you should ensure that you have a group of professionals around you for advice or referral purposes. These individuals or groups can help you make relevant buying and rehabbing decisions on an as-needed basis.

You don't necessarily need to hire these groups. You just need to do a pitch to each of them to see if they can be available to work with you on an as-needed basis or for an agreed-upon pay. Members to include in your team can come through referrals, networking, or local real estate association compilations.

These professionals include attorneys, contractors, accountants, realtors, mortgage brokers, hard and private money lenders, other investors, landscapers, demolition people, appraisers, bankers, wholesalers, bird dogs, cleaners, home inspectors, lead certification people and any other professionals you call upon for consultancy, work, or referrals. Having such a team ensures that your work proceeds steadily and without interruptions, in case you have buying, selling, or rehab questions.

Know Appropriate Time to Buy, Rehab, or Sell

Timing is everything in business, but more so in real estate. You may buy properties at bargain prices at certain periods. You may lock in contractors at reasonable rates at certain periods. Property selling may be slower and go at reduced prices during certain periods. You need to be aware of these periods so you know when to buy, rehab, or sell, as holding costs may apply if you finish rehabbing a house and have to hold on to it to wait for good selling periods. You also need to know these periods to plan the work on your rehab.

Outside projects like landscaping, walkways, roofing, and exterior painting may not be scheduled for winter seasons due to the weather. You may do this work during the right weather conditions and do the inside work during winter. Knowing when to schedule certain projects keeps the contractors busy throughout the rehab without interruptions, which may be costly due to holding period expenses or contractors asking to move to other projects or other competitors. I once made scheduling mistakes this way, and contractors left to work for other people. It took more than three months to get them back to my project.

The big takeaway here is that intimate knowledge of your project helps with the hiring, timing of work, and knowing which materials to buy and when. You will also know when to advertise for sales and rentals, the type of rehab to undertake, and in which neighborhoods to do a high or low-end rehab.

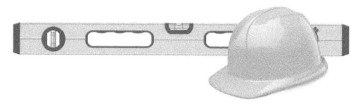

Chapter 6:

Shortlisting Only the Relevant Contractors

H ow do you hire the right person for your intended project? How do you know that the person you have hired is right for you and your project?

Allan, a senior partner at one of the local real estate investment companies, posed these questions to a group of investors he was teaching, as he does on the last Saturday of every month at his company offices in Baltimore, Maryland. He started his lecture by saying an investor must approach the contractor hiring process with the same rigor an employer goes through when seeking to hire a key employee. You must know who you want, what they will do once hired, when they will start, and when you want to finish the project. In other words, you must be familiar with your industry rules of engagement, the process of hiring, and the chemistry you are looking for with a contractor.

In many cases, you should break up the interview process into two or three segments. One part may consist of a meet and greet and completing a preliminary questionnaire. The questions listed at the end of this chapter should guide you. The potential contractors may opt to come to your office to fill out the questionnaire, you may email the questions to them, or deliver them to their worksite.

At best, encourage them to come to your office, if you have an established location. That way, you put a name to the face and you may start assessing their personality traits from the first-time impression. Dropping a questionnaire off to their worksite helps you achieve the same advantages as coming to your office, but also exposes you to his work environment. Some people may argue this is the best option since you will be able to see their worksite, enabling you to evaluate how organized they are at work, their seriousness, and even their employees, if they have any. You may also meet some of his employees or subcontractors. This gives you an idea of the crew he may bring to your workplace should you hire him.

Obtaining Contractor Names to Invite to Interview

Referrals can come from friends, neighbors, relatives,

co-workers, local real estate meeting groups, meet-ups, social media, churches, and many other forums. For many investors, referrals may be the vehicle of choice, since the contractors they refer are likely to have some proven track record. You still need to ask those referring some basic questions like the type of work the contractors did for them, how long they have known the contractor, were they timely arriving to work and in finishing the project. Were they cooperative and easy to work with? How did they handle being corrected? Were there any fights? Did he do what he promised to do, and did he honor the contract in general, if there was one? While these questions may seem like a lot, you can choose which ones to ask in case the person referring does not have enough time to talk or is uncomfortable being asked questions.

Don't forget to thank the person doing the referral even if they don't answer any of your questions. You will still need to do your due diligence on each contractor no matter how highly recommended they come to you. Some of the contractors who have benefitted me the most have come through reputable referrals. Unfortunately, some of the contractors who have ripped me off the most have also been those I have worked with in the past, or have come highly recommended through family members or close friends.

I have learned the hard way on many occasions. I realized that even though knowledge which comes the hard way is well-learned and retained, it's not necessary to go through hell when help is right around the corner, if you know where and how to look. Be that as it may, I was able to develop some policies that have worked for me over time. These policies have been applicable to other businesses as well. I learned not to let my guard down even when recruiting through family and friends, recruiting family members, or from a pool of those I have worked with, and even church members. I developed a good rule of thumb through all my tribulations: never hire someone you cannot fire. They will come to work for you knowing you will not have the guts to fire them, and so they may approach the job from a relaxed atmosphere. They may find it hard to accept constructive criticism. Overall, they may not treat your job with the professionalism it deserves.

Assuming you hire such people, please explain to them that business is business, and so the rules of the game as you apply them in your business remain the same for all folks, no excuse. You can explain that you can be friends or relatives outside of work, but at work, its business. It's serious business! Ensure they understand that and are willing to comply before you consider them for hire. Have them initial on the part of the contract that states this in your company policy. Due to my rough treatment at the hands of contractors, I have developed a strict shortlisting and hiring process for contractors. I have also learned not to rush in the shortlisting process. Since the contracting process is important and can eventually prove costly in the home improvement business if not done right, I have learned that it's best to wait until you find the right prospects

It's ultimately cheaper and less stressful than hiring quickly and having to deal with firing, arguments, potential lawsuits, and other forms of workplace chaos. Not to mention that the new incoming contractor may start from demoing some of the work of the prior contractor, before continuing. A new contractor you bring in to rectify the mess created by the outgoing one is likely to ask for money for re-doing the fired contractor's jobs. This will increase the overall rehabbing cost. The new contractor may also have a different vision than the prior one, and so confusion and increased costs may rule the day. You will avoid much headache when you are meticulous during the shortlisting and hiring process. As such, shortlisting and hiring is one of those instances when patience, actually extreme patience, may pay off big time.

Drive Around the Neighborhood

Sometimes the best way to meet a contractor is at their workplace. They are already familiar with the neighborhood and the styles that are in force. You also see them in action, and if you are lucky, they may tell you their story completely if you don't approach them as a potential employer. They may treat you according to the mood you find them in, and that may be good, as you want to evaluate them in different situations and settings. You just need to approach them right. You can try to approach them as someone who was driving by, saw the good work going on, and you thought you'd stop by to see the wonderful people making this happen.

The contractor might feel flattered and talk to you in a friendly manner. Your dressing needs to be professional casual, and don't visit in a sleek car. You don't want to appear snobbish. They should not feel threatened. Then the conversation will feel relaxed. Ask them if they can walk you around. Alternatively, if they don't have a problem with you walking around, you can do it by yourself. You could mention that you may buy a property like that in the future, and you may wish to talk to him. Whichever way the visit ends, don't give them a questionnaire on that day. You can do that on another day. You don't want to appear sneaky. Ask them if they can give you their business card or contact information in case you want to talk another day, or for referral.

Local Real Estate Agencies and Those on the Web

These may be on realtor websites, Zillow, Angie's List, Craigslist, Realtor.com, auction sites, Homeadvisor.com, et cetera. You can get contractor names through these sites or those associated with these websites. Whichever way you get the names, make sure you do your due diligence thoroughly as previously outlined.

The City

City agencies that deal with contractor issues may have some databases you can log into to get contractor names. For legal and conflict of interest reasons, city employees will not directly refer contractors to you, but you can get some referral information from visiting the rent courts, registration offices, or even a code violation. Property owners, agents, and even contractors do visit these places.

Building Supply Stores

It's not uncommon to see contractor fliers in home building store noticeboards. I have gotten contractor contact names in the Home Depot aisles for different projects. If you need a plumber, spend some time in the plumbing aisles, and you will bump into plumbers or people who could refer you to plumbers. Talk to plumbing associates. Use the same tactic for different projects like roofing, electricals, and flooring. For example, for an electrician, move around the electrical aisles. Talk to those you may see are looking at stuff like they are knowledgeable. Ask if they are electricians and whether you can get their contacts. If they are not, ask them for a referral.

Other Investors

Talk to other investors who may be doing what you want to do. You can meet them at the meet-ups, local real estate associations, networking parties, rent court, meetings, city agencies, attorney, and accountant offices.

Contractor Pre-Interview Questionnaire

A pre-interview questionnaire mentioned at the top of this chapter may look like this:

1. Name
2. Position applying for
3. How did you hear about us?
4. Employed or self-employed (Please include company name)
5. If the company is licensed, ask for the license number.
6. If the individual is licensed, ask for their license number.
7. Size of the company in terms of employees or assets under management
8. What's the contractor's title?
9. The company you work for
10. How many years with the company? How many years in this type of project?
11. How would you describe your level of experience related to the position you're looking for?

12. How many companies have you worked for or owned in the last ten years?

13. Please list any work references we may talk to.

14. Please state if any of the references are based on the type of work or position you are applying for.

15. Who handles the buying of materials whenever you take up a job? You or the owner?

16. Are you insured, and could you provide proof of insurance as part of the application process?

17. Does your insurance cover workman's comp and general liability? Do you agree to furnish us with the same coverage information from your subcontractors before hiring?

18. Do you have a portfolio of some of the projects you have worked on? Please share during our sit-down interview.

19. Do you have any bids in progress now? How many?

20. Are you always in charge of jobs at the sites?

21. Have you used subcontractors? Do you pay them directly?

22. Do you sign a release of liens or notarized affida-vits before requesting payments?

23. What are your average work hours per week on a project?

24. What methods of payments do you normally accept?

25. How much advance payment do you normally ask for? Is this negotiable?
26. Have you filed bankruptcy, and if so, what year and names—personal or business?

Every investor can pick and choose which questions to include for their specific hiring situations. If the job is big, you may use all the questions. If it's a small gig, use only those that are relevant to the project you are hiring for.

When Hiring a Friend Became a Nightmare

John learned the hard way when he hired a close friend of many years without shortlisting and taking her through the interview process. Mercy is a long-time investor and contractor who owns many properties and sometimes works as a contractor. In fact, the way she moved around criticizing other contractors' jobs and saying how best she and her team work, John always wanted to give her one of his projects, but their timing was always off. When John had a job, Mercy was busy.

This time around, when John had just fired another contractor for various contractual misdeeds including no call, no show many times, shoddy work, and purchasing sub-standard materials, he approached Mercy to take over the work. She accepted on the condition that it was after she finished her on-going own work in another two months. John accepted. "I am so honored to work with you. I have always hoped for this occasion."

"No, no. The honor is indeed mine. I like when I work with a true friend like you entrusting me with a job. We shall get this thing squared away."

John was super happy. This was a person he had always wanted to give his projects to, given how highly she talked about herself and her work. To John, this was like one of those dream come true scenarios. He had always looked up to Mercy as a role model. She had been in real estate for a long time and didn't seem to have these horror stories that other investors had with contractors. Or so he thought. She had a loyal work crew, as she always stated. John had never met them in action nor visited any of her workplaces. He and Mercy talked many times on the phone, comparing notes as investors. She always projected this air of a no-nonsense person, who had it together with contractors. In fact, she always advised John how to work with contractors.

When this opportunity to work together came, John didn't think twice. It was like working with his role model. He never bothered to look elsewhere. He threw to the birds his newly acquired knowledge on shortlisting or interviewing contractors meticulously before hiring. "How do you interview a person you already believe is an expert in her field, who owns more properties than you, and has been in the business ten years longer than you?" John told me this story one day. He recalled how humbled he was at the start of their work. "I was prepared to learn, to be mentored, and even to be blamed for not sticking to my rules of working with contractors." He was willing to take anything that was going to be thrown his way, so long as his project was straightened out and finished.

Two months went by. Mercy finished work on her property. As they had agreed, she and John met at the worksite. She didn't disappoint. She ripped the project apart. As John anticipated, she used all the expletives in the world to tell John how much he had dropped the ball with this contractor. As someone who never minces words, she gave it to John, telling him how costly it will be to repair the project. She said to John it was like he had let this contractor defecate all over him and then he thanked him with a bundle of dollars. This is crazy. John kept his eyes on the prize of her getting the job done. All other utterances, no matter how stinging, were basically noise.

When Mercy talks, you have no choice but listen. She's a six feet five inch-towering personality. She's tall, dark, bulky and loud. She's fond of saying that God gave her the physique to operate in a man's world and that's why she's comfortable among the contractors, kicking their-you-know-what when they try to flash their stupidity in front of her. She says many contractors are bullies, so you have to know their game plan and play better and above them. She believes that's why contractors respect her. They know she's always a step ahead of them.

She also says she has no problem invoking her femininity to win an argument. For example, when talking matters of design, she wins her arguments by saying "from my point of view as a woman who knows what we appreciate as women in a home, this is how the kitchen will be done." She has learned that typically men back off when she invokes her 'woman-ness' as she puts it. Mercy was always colorful whenever she and John talked on the phone before they started work. In one such phone call, she laughingly told John that one of her employees is fond of telling her other employees that when he gets a chance to give it to her, he fights her like a man because she's a bully who doesn't deserve courtesies extended to females. This employee believes Mercy is a lion who will throw you out of the den if you don't throw her out first. Mercy believes this guy is one of her best employees because he is a hard worker who tells it like it is. She says they see the world the same way, "outshine or be outshined…and so our chemistry is in sync. That's why we work so well together." It's all in good humor, she adds. They have an understanding that if it helps each of them to fight tooth and nail to relieve whatever stress is in their chest, so be it, so long as neither of them holds a grudge. They fight like dogs, then let go and move on with the job at hand.

John adored Mercy, at least up until that time! He believed working together could easily pierce their relationship, and he hated to think of that. Mercy drew up a contract, which John did not scrutinize properly before signing because he was dealing with a long-time friend. Big mistake. To make matters more confusing for John, Mercy said she was not going to take upfront advance. "John, you are my friend, I don't need to take money from you before I start work. I trust you," Mercy said at contract signing. "I want to show you that I'm not like any of those contractors who take money from you, and not show up. I will do a great job for you. You will see." John was beyond elated.

Mercy started work as promised. She bought materials with her money. She said she would present receipts for reimbursement every week. She came in for three consecutive days with her crew and started work. John could not figure this out. Someone working with her crew without asking you for money to pay them, and even buying materials with her money. This was beyond an act of kindness! Nobody does that. John wondered what the catch was! He shared his good news and bewilderment with some investors. They all couldn't understand what this good friendship was all about. Some of them had been in real estate much longer as investors, and they never came across this good fortune. Everyone believed there was a catch. They just could not lay their hands on it. They said this lady was up to something. One friend even said this is unreal and fishy in this industry where many people are hyenas, rude, and some are outright crooks and thieves. Everyone settled on a wait and see attitude!

Mercy worked on this project for three days, took one week off. She said she had emergencies to attend to in her projects. She said she expected John to understand her since she's not just a contractor, but an investor also, having her projects to work on. "I wear the two hats well - don't you worry!" Two weeks went by and she was still a no show. She came back on the third week, and worked well through the following week. Then she took another week off to go on vacation. "In this business, you have to force yourself to go on vacation, otherwise you go bananas. Everything including the contractors, tenants and the city keep pushing you to the edge." She told John over the phone one night. "I found myself sliding toward the edge. I had to apply sharp brakes. That's why I went on vacation abruptly. Forgive me, I will be there next week." She came back as promised, worked for three days then went to work at her property saying she had another emergency.

John was beginning to be mad as he later told me, "I started to develop jitters. I wasn't seeing what she had promised. She was just like the other contractors, even worse. She was off many more times at the beginning of a project than I had ever experienced."

John decided to confront her with these facts, telling her what was truly in his mind. He knew he was going to say something. He just didn't know when. That time had come sooner than he thought. She was missing work more than he could stomach.

Not someone to take being challenged lying down, she retorted that the difference between her and those other contractors is she had not taken a dime from him. "Now I realized her idea of not taking advance money from me was a way to shut me up if I complained." He told me one day. She had also told him as much. "How dare you talk to me that way when you haven't paid me a dime?" John told her he has always wanted to pay her from day one, but she told him to hold on. Now she was using that as an excuse as to why he should not correct her. She assured John she would finish the project on time. John did not want to bring in another contractor, so he went along with what she said but with major reservations. Soon, she started to ask for money. Problem is, when she asked for it, she asked like she needed it yesterday. She asked like she almost expected John to drop everything he was doing and rush to give her a check on the spot when all along she was the one telling John to hold off.

John told her they needed to be organized and follow the contract, including payment schedules. She said she's not an organized personality and claimed John knew that before hiring her. "I like to fly by the seat of my pants. I have done well for myself that way in real estate. This is not office job. No need to change this late in my career, my friend." She also said if she had known that John was such an unappreciative person, she would not have taken the job. "It's keeping me from working on my projects." In the desire to finish the job, she brought in other contractors who not only didn't seem to know what they were doing, but drank and did drugs right in front of her. When John complained, she said these people are drug and alcohol addicts who could not function without them. She argued that it is the reason she continues to tolerate them. They are loyal and dependable, but they have vices like everyone else.

At one point when John inspected the work, he found that none of the door frames had been cut and repositioned in a straight way. He complained to Mercy that this is a result of alcohol and drugs. Other things were also getting messed up. The laminate flooring had been laid so poorly. There were spaces between the laminates. The edges were crooked, and the transitions were off. On the roof, one contractor had coded another homeowner's chimney thinking it was John's. All the plumbing in the house was either leaking or loosely fitted. The electrical wiring was another mess. Some wires in the basement were left hanging all over the place. This new crew had to repeat things three or four times to get right. In many cases, like the flooring, John had to bring in other contractors to rectify the situation.

It all became too much for John to bear. So much for the friendship, he thought. He decided to let her go too. Another bad experience in the books. Even the type of people Mercy employed or hired temporarily were despicable. If any of these druggies and drunkards got injured on the job, they would probably find a way to go after him. John and Mercy negotiated out of the contract. He was staring at a second loss in the same house. It was a mess. He decided to take a break from this house, work on his other houses, take stock of what he learned, then come back and finish.

He learned many things with this work with Mercy. Most of these lessons, he was already aware of. First and foremost. Set rules for hiring and management, and follow them whether you are dealing with a family member, friend, pastor or someone who makes you believe they are experts. Even for experts, the way someone runs their business is not the way they will run yours. When hiring, follow your rules for shortlisting, interviewing, and hiring. You need to see their past or current work if you are the hiring authority. Don't believe what you hear from anyone. Verify, verify. When someone wants to work for you without accepting payment as is practiced in the industry, be extremely careful. Free is never free.

There is always a catch. Go by what your contract says. It's not easy to fire someone you have not paid. They will always tell you that they are good to you and you are not reciprocating, even if you have money to pay them, and so they are not doing you any favors. Finally, the rules for letting go of a contractor should apply to all—friend, family, or foe. Don't let mistakes linger.

In general, at the shortlisting stage, the key to putting together a good prospect compilation is through referrals, networking, and talking to other professionals who have done similar projects to yours. You need to have a good idea of what you want and to know how to approach hiring the right talent who has the expertise you desire.

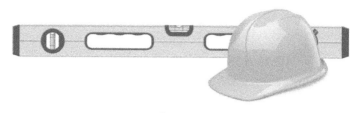

Chapter 7:

Getting the Interviewing Process Right

The preparations you did at the shortlisting stage should make determining who to invite for interview easier. The pre-interview questionnaire that a prospect filled out should give you ample material to weed through to select those contractors likely to get the job. As the pre-interview questionnaire is long, you want to use some key points in there to guide your decision on who to invite. These are the references, certifications, relevant adequate experience, payment structure, and personal reliability.

For references, you are looking for at least three past clients. If they have not provided the names and contact info yet, you can call, email, or text the contractor to provide no less than three past clients you can contact. The more the better, but three should suffice. On qualifications and certifications, you want to make sure they are licensed and insured. They should have provided their license numbers and documents showing they have general liability and workman's comp insurance. You can communicate with them to send the documents if they have not done so. Your goal is to confirm that the contractors you interview have the certifications and qualifications appropriate for the work you will be hiring them for.

In terms of adequate experience for your job, you want to look at the relevant project management time under their belt for specific work you are considering them for. Your payment structure or schedule should be in sync with how they want to be paid. For example, if you have taken a rehab loan, some lenders have strict draw timelines and the type of inspections that trigger these payments. You want to make sure the contractor you hire can work with these lender guidelines, otherwise, there will be constant arguments and mistrust between you and the contractor.

Finally, you want to ensure that the contractor you hire is reliable and trustworthy. You will peruse the pre-interview questionnaire to look at the elements that can show you that the contractor is reliable and keeps his promises. You can also get these from the referrals.

Also, depending on the size of the company, you can get a history of complaints on them from the Better Business Bureau, and state and city agencies. The local home builders association could help you know if their business license is current.

These five factors above will eliminate many of the contractors that do not meet your minimum qualifications. The hope is that you still have enough applicants to choose from. If you don't have any contractors left in the pool, then re-do the shortlisting process. Hiring contractors is a major undertaking in the rehab process, so don't rush it or choose from a wrong pool. It is part of the hiring process you definitely have to get right to have a shot at selecting a good candidate. So if you have to re-do the process a couple of times, it's worth the wait. You will save money by patiently waiting to hire the right contractor for your job instead of choosing one who may mess up your job causing you to pay a new contractor to rectify the prior contractor's mess.

When you are ready with the final list of contractors to invite for an interview, review their information again to see that your final listing has the best candidates to choose from. Double-check that the contractors you have on your list are only those who have done the relevant type of work you are interviewing for. Zero-in on their experience and the size of projects they have handled. You are looking for those who have handled projects matching the size of rehab you want to work on and above. The bigger, the better.

For those you are highly considering, ask them if you can now visit their worksites officially. Clearly inform them that this is part of your interview process. If they show some resistance, inform them that your lenders require you to visit them as part of the funding consideration. Many lenders I have worked with ask to know the type of due diligence you have performed on the contractors you've hired. Be sure to also put in your file the pictures they provide of the past projects they have worked on.

Once again, review their licenses, certification, and insurance. You can ask the contractors for any clarifications they may have at this point. Mention to the contractors that if they get the job, you would like them to list you as "additional insured," just in case any claims may spill over to you. Explain that the proposal they will give you should show a firm timeline for starting and finishing the project. Tell them that a large penalty will be put in the contract if a timeline is missed. A large enough penalty makes the contractor pay attention to finishing times. Also, explain that part of the interview will involve providing a detailed written estimate and contract for the work. In the contract, they will state that any changes to the agreement or deviation from the plan will need to be done in writing and agreed-upon by both parties. Finally, tell them that they will need to state how much advance payment they will be asking for.

Time to Ask for Contractor Estimates

At this point, you are relatively comfortable with the prospective contractors you are still working with. You have done your homework and consulted on the scope and cost, and have a good idea of whether you can afford to finance your project. It's now time to contact some contractors for estimates. You can share with them the blueprint and scope you have developed. If you have made other decisions, like materials to use and their costs, and what you want the final product to look like, share that with them so everyone will be working from the same script in arriving at estimates.

Ask them to provide a detailed proposal. You are hoping for at least three proposals. The more the better, but three should be the minimum you want to work with. Let the contractors break out the jobs by price and timeline in their portfolio. They should include their proposed draw schedule, which should have room for negotiation and flexibility. Please tell them to state who will be responsible for trash removal. You want the contractor to be responsible for trash removal, even if that increases their contractual amount by a little more. You want the work area to be clean as they go, and so letting them be responsible for trash removal will force them to keep the site clean. Make sure to discuss cleanliness as part of the ratified contractual agreement. Once you get the proposal and estimates, add about ten to twenty percent for unexpected and unforeseen costs to see whether you have enough resources to hire a contractor and finish the project.

With all the above in place, invite the contractors for a sit-down interview to go over the information they have supplied and those that you have gathered. On the day of the interview, start by going over the job you have, stating the job expectations and the timeline for completion. Collect copies of any certifications and documents you asked for which they have not submitted. Ask for their government ID, check DUI, driver suspensions, and any other issues you believe you should know. Go over their work history, making notes on any inconsistencies or long breaks on their work experience timeline. Ask for an explanation of these inconsistencies. Good contractors don't have long gaps between jobs. Too many gaps could portray poor planning skills, and these may be a reflection on how they will handle your project.

Take note of the individual's demeanor and presentation. Remember, you need to evaluate chemistry between you and him. Rehab projects can take a long time, and so both of you will be working together for some time. You don't want to be bound together in a project with someone you may feel uncomfortable working with. Also, complications may arise, and you want someone with whom you can calmly resolve problems. These complications may include property failing inspections, the weather, and other unforeseen delays. At the end of the interview, inform the contractors when you will get back to them with a decision. Also, ask for permission to call them for any clarifications, should you have any. Extend the same courtesy to them so they feel free to contact you at any time with questions.

At the end of the sit-down interview, walk them to their truck. Mention how well you like the truck and that you would like to see how the inside looks. Your goal is to see how organized they are. Remember this is part of the interview. A contractor whose truck is organized is likely to keep your workplace organized.

When investors ask the right questions, get the needed documents, are observant during the interview, and do proper due diligence on the candidates, they will hire the most suitable contractors for their jobs.

Chapter 8:

Evaluate, Eliminate, Hire, and Even Fire

Hiring a contractor is one of the most important jobs an investor will undertake in a property rehabbing process. This process requires patience, knowledge of what to look for, and how to evaluate prospects and hire the most suitable ones for your job. This meticulous selection can be fraught with challenges if not handled well. Hiring success requires proceeding methodically, starting with aspects that may be obvious and important such as an individual possessing the city required certifications for your type of job.

Another red flag to look for is if the contractor's references do not return your calls within a specified timeframe. You may summarily rule out those candidates.

Your reasoning may be that, if even the people he has given as his references are not responding well or flatly refuse to respond, then something may not be adding up. Don't get involved with such a contractor. The process of contractor selection after interviewing should follow your criteria with minor deviations if necessary, depending on specific circumstances.

Time under project management in a similar job is a straightforward evaluation criterion. They either have the necessary experience or not. For licenses, they either have them or not. No middle ground. For payment schedules, this can be negotiated if all other aspects are lining up, so long as you are in control of your finances or you can convince the financiers to work with you. Reliability or unreliability is also straightforward. You can get insight on these from the references, or from the way the contractor presents himself at the interview stage. Trust your instincts.

You will also realize when evaluating many aspects that most of the dots do not always line up as you want in your carefully listed criteria. You will be extremely lucky if any particular candidate meets all your requirements. This is where knowledge and experience come in handy. You need to know which factors are not negotiable and which are flexible. Just make sure the most important factors to you line up. For example, if you want a high-end rehab, you are best served if your key criteria involve someone who has had experience with high-end stuff. The more years of experience under their belt, the better. The reason is that a high-end rehab involves high dollars, and a small mistake could be quite costly. Also, the buyers in this group tend to be particular and do have keen eyes.

You will need to evaluate the contractor's timeliness for arrival for the interview. When you start with someone coming late on this important day, you can be sure that this is the way they will show up at your project. Late! The next item to check is how organized they were during the interview with their documentation.

How about their truck or actual job site where they currently work? Clean and organized, within reason, or untidy all the way around? Contract work can be messy, so use discretion when judging this aspect. Experience shows that contractors who are organized at the job sites and are organized in other aspects of record-keeping will do clean projects and finish in a timely manner.

Ease and willingness to readily produce their contractor license, insurance, and bonding documents is another important evaluation moment. When they argue, reluctantly produce the documents, or keep asking why you need them despite your explanations, then that is not a good sign. You can take it to the bank that they will drag their feet in various aspects after they start working for you. You can believe that that's their nature or they are trying to hide something. Either of these are red flags that you should not hire such contractors.

The proposal they provide is another area you want to pay attention to. Providing reasonable estimates is key in this evaluation. You already have your budget estimates after doing the scope of work. Too high or too low a proposal can be a red flag, signaling that a contractor may not know the job, is greedy, is trying to find a way to get in then jerk up the price, or is attempting to see what he can get away with. Rehabbing is costly and time-consuming, and so you don't want to entertain any form of games. Other aspects to consider and eliminate outright are those of a contractor not possessing the necessary and relevant experience in the desired area. If a contractor only has timeframe issues, then you could keep him, and see if you and the lender can work something out in case he is the most qualified person you have interviewed for this project. If his timeline is way out of sync with yours and negotiations are not possible, then there is no need to consider him. You can put him down for your next project.

Also, if a contractor is doing just a portion of the whole work, like plumbing, and he is working under another contractor who is licensed, verify who the master contractor is and that he, the master, is aware and agrees to this setup. In such arrangements, confirm who you will make payments to—the master or the one doing the work. If you are unable to verify this arrangement in writing with the master plumber, then you should not hire the contractor. You don't want to get entangled in arrangements or agreements between parties that you were not part of. I once hired such a plumber who pulled a permit in the name of the master plumber. The work was poorly done and I approached the master plumber, who informed me that he was not aware that a permit had been pulled. Since the permit was pulled in his name, he agreed to rectify the problem and call for a city inspection, but only on the condition that I paid him double what the earlier plumber had charged me. I was faced with an impossible situation, since no other contractor would continue a job where another permitted work was pending after a permit had been pulled. They would need to pull their permit and charge me to undo the earlier contractor's job and then charge me for doing theirs.

After discussions with different plumbers, I realized I was between a rock and a hard place. The cheapest option was to pay the master plumber the double amount he was asking. I obliged. I paid the master plumber and then promptly fired the plumber who was supposedly working under the master plumber. He had lied to me. I learned a painful lesson with this hiring. Be careful if you have to hire a contractor who mentions that they are working under another person's license. Best not to be

- Your intuition and instincts tell you that you may not trust the contractor alone with you in your home.
- No proper license, no registration, no adequate general liability insurance, and no worker's comp insurance.
- Doesn't possess relevant skills to your particular project.
- Comes late to your appointment or is a no show at all.
- No proper relevant tools.
- Asking for large payments up front, or only want cash payments.
- Insists that you hire him immediately, as he just finished another job. Explain to him that hiring is a process that you have to follow. Don't succumb to high-pressure tactics. Good contractors understand they have to give you time to make proper decisions. On the ones who rush you, you need to ask yourself: what do they fear? Why are they in such a hurry, what are they running from?
- Doesn't seem to communicate well, like someone knowledgeable about your project would.
- Seems to display a bad attitude, or is rude. Both could signal you may have communication challenges between you. If it's a long project, there can be communication pain for a long time.

This list has been helpful to me as a quick checklist on who not to hire. However, as with many aspects of life, this is mainly a guide, and specific situations may demand different treatment.

Experience and knowledge will be your best driver. For example, a new contractor may be preferable to a seasoned contractor with no proper history and no references. If you can't get a good contractor with experience, then you could hire a new contractor who interviews well. You could give him a small job for a test to see how he performs before you decide if you can take a chance on him with a bigger project. In general, if a contractor is licensed, bonded, in good standing, and many of the above criteria check out well, then consider hiring.

When to Pull the Trigger and Let the Contractor Go

You have hired this contractor. You have established a great working relationship. You are buddies. He shows up to work on time and even stays late. By all intents and purposes, he is a good guy. Problem is, he is a lousy worker. He has to go over jobs two or three times to get right. This increases time to finish, and stretches the budget. You have talked and talked. He understands and acknowledges his shortcoming, but there is no improvement. He's a bad worker, period. What do you do?

Jerry, a relative, came to me in June 2016. For once, I could tell everything was not copacetic. She's not a person who wears regret or failure on her sleeve. She's a bubbly personality, and talks softly but with authority. When Jerry is in your presence, you listen, and you listen attentively. She has a way of looking into your eyes and assuring you with her hand gestures that you are buddies working toward a common cause. She can turn an enemy into a friend on a whim. She's firm and persuasive.

She's also one of those personalities who can hide her feelings, so you may not know whether you have offended her or not. It takes a long time to know if Jerry is defeated. She does not admit to failure that easily. She has had a series of successes. Failure is not generally an option for her.

I have known Jerry for many years, both in business and casual settings. She's a go-getter and one of the most persistent personalities you will ever meet. When she admits to a defeat, you know she's defeated, and you have to take her seriously! This idea of continuing to push despite constant failure is one of her great strengths, but it's also one of her greatest un-doings.

Due to the high dollar amounts involved and government regulations to adhere to, real estate rehabbing is one of those areas where you cannot keep sugarcoating situations when they start to go bad. You need to correct situations as they occur, within reason, or you will pay dearly with your wallet.

Since Jerry has more successes in life than failures, I can count instances when she has failed. On this Sunday, in June 2016, I came back from church around midday and saw a car that resembled hers parked in my block. I parked my car at my house, then walked toward the car. It was hers, but she was on the phone. I knocked on the side window. She motioned me with her hands that she will come. She's official in her dealings, and not one to show up at anybody's house unannounced.

As I walked away to my house, I decided not to go in. I knew something was amiss. Soon, I saw her leave her car and walk in my direction. As she came closer, her phone rang again, and she walked back. Then I heard her yell saying enough is enough, then she disconnected the call. The look on her face said it all. There was trouble in paradise, so to speak.

Not one to show emotions openly, she sighed and shook her head as she approached me: "This guy is costing me a fortune in time and money. What should I do?" She said as she pulled me away from the house to tell me her predicament.

"My brother," she said. "I need your help. I am in a bind." I looked her directly in the eye, and she looked down, uncharacteristically of her.

"I have been working with the contractor, a pleasant guy for the last three months. We have been working in this house that should have been finished this month, yet we are not even one-fourth of the way there. Martin, the contractor is a pleasant person, respectful, communicates well, shows up to work on time, and stays late. Problem is, he takes one step forward and two steps backward. Every time I visit the property, we appear to be moving backward."

"How so?" I asked in a low, subdued tone to be respectful of the situation.

"I think it's time to cut him loose, but I'm conflicted since he's a nice guy. We relate well, but we are more than behind on this project. Please help me, what should I do? A house that should have been finished this month could easily take another six months the way this guy is working. I have hired and fired contractors in the past, but this is a tough one. He's so nice. It's easy for me to fire a bully or someone who is rude and doesn't show up to work daily and on time. Martin is the complete opposite.

"He is punctual, respectful and trustworthy, but work on the project isn't moving."

By this time, Jerry was beginning to relax and even look at me directly in the eyes. I could tell she had said most of what was in her heart. She had accepted the reality that she needed to do something. She just needed someone to convince her that firing was the right action.

I started talking by letting her know that being nice and trustworthy are great attributes, but they don't pay bills. I asked her the type of help she was seeking from me.

"My brother, I just want to finish this house. I am defeated. I hate to have another contractor come on top of another contractor. They will just blame the crap out of the previous one. They will even blame me. Then they will tell me that they will not be responsible if the work doesn't end well since they didn't start it. They will tell me everything that covers their behind and leaves mine bare. Sometimes, I get tired of contractors! You know them well. I don't need to tell you anything about contractors. You are the master of the bad, the ugly, and the beautiful!" she said to bring humor to the situation, and continued:

"You can hire the greatest contractor in the world, but the problem comes when they make mistakes. They rarely admit it. This is one of the strengths of Martin. He admits to mistakes, but then I have to buy new material for him to re-do his mess. In addition, you factor in the holding costs for the delay in finishing the project and your budget increases. Sad!"

"My brother, this guy has to go! Please help convince me that I am doing the right thing, and explain how you think I should go about firing him. How do I break the news to such a lovely person?"

"Well," I said, "first things first. We are on the same page that Martin has to go. Your predicament is how. I got you. I got you!" I assured Jerry. "Your question principally is: how do you fire a good person, and how do you communicate that?"

"Exactly!" she said.

"Let me start by sharing and paraphrasing one of the quotes from Mad Money on CNBC. Jim Cramer, the host, says: Some people want to make friends...I just want to make you money. You and I trade stocks in our other lives, so this is something you can identify with," I told Jerry and continued.

"In business, one of your main goals is to make money. You may have other goals, but making money is a key one. Friendships and all that are important, but you are not in business as a social event. To put it to you straight, in my opinion, when in business, you must differentiate between those friendships that make you money and those that lead to loss of money. Again, as we know in our stock trading philosophy, you let the winners run, and cut short the losers. In many cases, if I remember well from business school, many business decisions need to be made quickly. You cannot be good to others when you are personally bleeding. You have to apply the cure. In certain instances, firing is the cure. Starting afresh is a cure. Taking a break is a cure."

Jerry listened attentively as I outlined the background from which we will make the correct decision that sorts out her issue. You could tell from her face that she was not only confused, but in pain, ready to let go but also wanting to hold on to Martin. In my mind, this was more of a psychological decision than a business one. The business part was easy. She's lost money due to his repeated mistakes. The psychological part was how to part ways with a trustworthy and nice person. My job as I saw it was to lay out the business process of making a firing decision and how to methodically deliver the verdict that results in an amicable split, and lets the contractor move on with dignity.

"Jerry, as I see it, I can help you with the whole process. You need to remember that beyond all the friendship and niceties, you are in a contractual relationship. Let's lay out the business decision making process."

First and foremost, below are some of the reasons an investor may pull the trigger on a contractor:

- Poor workmanship affecting job quality, and the contractor has no expertise to repair or is unwilling to repair. You must have talked to the contractor, given him time to correct the anomaly within a specified time, and if there's still no progress, then let him go.
- Constantly behind in projects, and not honoring timeframes.
- Unilaterally and constantly exceeding contract costs without approval or having a written agreement authorizing such excesses. The

contract should state that any labor or materials cost increases must be discussed and agreed to in writing, and that such changes become part of the contract.

- The contractor becomes a bully, stops listening to you, and constantly violates the contract in letter and spirit without caring.

Document violations with video, texts, photos, involving third parties in meetings and conflict resolution, progress reports, and emails so you have evidence in case you have to go to court to defend the firing.

"Jerry, your problem falls in the first bulleted point. Martin has poor workmanship that affects the quality of the performance. He has no expertise to repair or when he does repair, he has to do so over and over again, causing you monetary loss in terms of materials having to be re-purchased and time. If you finished the house on time, you could start getting some return on your investment sooner rather than later. He is constantly behind, basically bursting contract costs."

"In my opinion, this is clear cut. The first thing we need to look at is the contract between you and Martin. Your contract should have a termination clause that should outline the steps to go through before legally pulling the trigger. By now, you should have the necessary documentation in terms of texts, messages, letters, pictures and videos, receipts, and all that. Once you have this documentation, and you have a back and forth in communication of what's going on, then the next process is pulling the trigger."

"Since Martin is not only a friend, but a trusted one at that, you can approach the process in different ways. If you have another job for which he is qualified and can do well, talk about transferring him to that job. Explain the reasons why. If you can, you may politely explain why you need to part ways. You can tell him that your financiers are on you to finish the project, and they will not give you any more money unless you switch contractors. You can mention that the decision is not in your hands, but you will continually look out for him for other opportunities."

"If you still don't have the guts to talk to him directly, then I can come in as the financier and blame both of you for delaying the project. I can come in when both of you are there and rip the project apart. Then I will give an ultimatum that I am cutting off the finances unless I see a new contractor. I can then let both of you talk privately to see if you can amicably part ways. If you can't, then I could write you a letter discontinuing your funding based on the project being behind. You can use this letter as the basis for terminating his services, quoting your contract termination clause that highlights the bullet points above."

As I explained all these, Jerry seemed to relax some more. She wanted my help with terminating Martin's services, hiring the next contractor, and helping manage the process to completion. She said this project has sucked her blood more than any she has done. I told her that real estate rehabbing is a tough job, with many moving parts. Everything needs to be planned and executed methodically. Managing contractors is one of the toughest parts in a rehab if the hiring was not done correctly.

I worked with Jerry on the firing process. I came in as the financier inspecting how my investment money was being applied. Since Jerry had told me that Martin is a pleasant personality, he was easy to talk to. I developed a relationship with him. I was there every day for a week. I looked at how he worked, and I explained that the project is so behind that we have to change contractors. I mentioned that I have taken over the supervision of the project, and so he will be dealing with both Jerry and me.

Jerry and I drafted a termination letter, which I delivered to him by the end of the week. I mentioned that we will keep him in our mind and will offer him appropriate opportunities as they show up. He complied and left without complaint. I helped Jerry hire the next contractor and also helped her manage the rest of the rehab process. She was so happy and compensated me handsomely—beyond what I had asked for.

Before any firing, ensure that you are pin-pointing the reason in the contract why it has become necessary to terminate someone's services. Such reasons may include

Chapter 9:

Killer Contract: Lying Contractors Beware

A t this stage, you have evaluated and hired a contractor of your choice. Even if they were not your first choice, at least you've chosen one of your best options among the applicants. You have to do the best with who you got. As you have read and learned in the prior chapters, rehabbing is a serious high dollar business which has to be treated with the seriousness it deserves. That means leaving nothing to chance. You have to follow your rules, rules of the rehabbing game you have mastered or hired a consultant to work with you and help you master.

With the contractor search settled, the next phase is the contract. Typically, the contractor draws out a contract outlining what he understands will be his responsibilities and expectations in the project. He presents the contract to you. You review and, if necessary, propose your changes. You both discuss the changes and come to an understanding of what your individual and collective responsibilities and expectations are. You have to agree on every tiny detail including finishing dates, materials, and all payment schedules.

Specifics to Include In a Contract

Once you settle on which contractor to hire, think about having a watertight contract that will help with smooth operations throughout the project. A contract lays out the operational parameters, so the investor and contractor are on the same page for the starting point, process, and ending. A contract can be any form of agreement that the parties agree to. Many of the good ones include the following.

Description

Scope of work, materials to use from beginning to end of project, materials quality and brand. Blueprints, plans, and any other drawings should be attached to become part of the contract.

Permits, Licenses, and Inspections

Specific information like license numbers and all

particulars should be included in the contract. It's the responsibility of the contractor to provide them.

Timeline

Completion timelines to be spelled out clearly. These timelines will guide payment draws and final payment dates. Daily start and end times should be specified, even though the contractor may be operating independently.

Penalty

Agree on the dollar amount of daily penalties for any day the project is late.

Payment

In many cases, do not pay more than $10,000 down, and basically, no more than one-third of a project's total cost. Check with your state and city if there are laws that regulate the amount of down payments. Determine and clearly define project completion phases and how much will be paid upon such completion. For example, some people will pay when electrical wiring is complete and passes inspection. Same with plumbing. Some people will pay per floor completion. Stick to what you agree on.

Insurance and Property Damage Liability

Clearly state who will be responsible for damage to the property during rehab. Ensure contractors have adequate general liability insurance.

Warranties or Guarantees

Should cover materials that contractors buy and the total project work covered. A contractor who does not provide a warranty or guaranty may not be good since they cannot stand behind their work.

Lien Waivers or Lien Releases

Tell your contractor that you will need a lien release or waiver at the end of the project before the final payment is made. The lien release should state that any supplier or subcontractor owed any money, by the contractor, is the sole responsibility of the contractor, and not the investor.

Given that a contract is an extremely important agreement binding the parties, it may be a good idea to have another real estate expert, or better yet, a lawyer to look it over. The goal is to make sure it's broad enough and in-depth enough to include all areas a legally binding contract covers, such as:

- Contractor to produce an extremely detailed and written contract showing specific figures and key project finishing dates.
- The contract should be extremely detailed to include, for example, types of roofing, plumbing, types of materials, electrical fixtures, decorations, et cetera.

- Any advance payments should be included in the contract. A payment draw schedule should be included for various task completions. The method or mode of payments should be spelled out. If materials are included in the payment, clearly state which materials are included with the total payments. For everyone to be on the same page, remember to describe the quality of the materials to use and if possible, state the name brands.

- Pulling of necessary permits should be spelled out, and by whom. Permitted work should only be paid after passing inspection.

- State that all subcontractors are the responsibility of the contractor. Make sure they have been paid in full before making the final payment to the contractor. Subcontractors should also be licensed and certified

- The contract needs to have a termination clause for cause. Such cause may include lateness, shoddy work, job not done to code, substandard materials not approved by owner.

- Workflow and payment. Only pay with a check, credit card, or other modes that leave a trail unless there is a prior agreement and arrangement.

- State in the contract who is responsible for trash removal.

- Make sure the contract has a dispute resolution clause(s).
- State that work will not start until the contract is signed and all the agreements are documented and in place.

Before signing, ensure all the necessary paperwork that should be attached to the contract are in place. These are, but may not be limited to, independent contractor forms, scope of work, sketch drawings if any, building designer and architect forms if any, payment structure, certifications and insurance forms, W-9 tax forms, and lien release or waiver forms.

Independent Contractor Agreement

This form details everything about the project and the price and also defines the relationship as different from an employee/employer situation. The contractor will set his times of work which will be communicated to the business owner/investor and included in the contract. Basically, he is his own boss, exercising control and independence, although within the written contract and agreement. He and his team will also not partake in the employer benefits. He is not an employee of the investor.

Scope of Work

This write-up details the full extent of the project,

including materials to be used, sections to be demoed, type of bathrooms and kitchen, and everything else in the property that will be worked on to the tiny details that can be explained.

Payment Structure

This will outline when periodic draws and final payments will be made, typically upon the passing of the city inspections for permitted work, or owner or financier inspection as agreed to on in the contract.

Certification and Insurance Forms

As already explained extensively elsewhere in this book, these include contractor and business license forms, general and liability insurance, and workman's compensation forms. You should demand and get updated forms showing that the contractor maintains adequate insurance for himself and his workers throughout the project.

W-9 Tax Form

This is a required IRS document to be filled out by independent contractors.

Final Lien Form and Waiver

This form is used at the end of the project, but you should mention to the main contractor to get one for you and himself anytime a subcontractor finishes whatever he contracted them to do. The contractor should be giving you these forms for your file as soon as the subcontractor signs them. The main contractor will sign his own when the whole project ends. These forms for the subcontractor give you, the investor, a waiver from a mechanic's lien that the subcontractors may file against the property if the main contractor has not paid them in full. It is important to note that even though these subcontractors are not directly hired by you, and you don't pay them directly, the law allows them to file a lien against your property if the contractor who hired them has not paid. The lien form which the contractor signs at the end of the project similarly protects your property from a mechanic's lien (more detailed discussion of mechanic's lien is below).

Once all the forms are prepared and everything is in order, you the investor should schedule a meeting and walk through the property with all the parties involved (contractors, subcontractors) and review everything. This gives everyone peace of mind and puts everyone on the same page with regard to the project details, timeframe, and budget. Any confusion should be addressed immediately. With all the parties in agreement, proceed to sign, date, and ratify the contract, making sure copies are given to all concerned.

How a Mechanic's Lien Functions

A contractor can file a mechanic's (construction)

lien with a local jurisdiction if he believes you have not paid him for services rendered per your agreement with him. The lien makes your home become security for outstanding debt on services or materials which the contractor claims is due and unpaid to him. After some time, according to your state law, the contractor could sue you to attempt to collect this debt by forcing a sale of your home.

A contractor may file this lien for something as simple as a minor payment you refuse to make because the door cabinets do not close, or that he put a different faucet than the one described in your contract. While the specifics of what creates a valid mechanic's lien varies from state to state according to individual state laws, you need to understand that this is a lien you may have to deal with before you can sell or refinance your home. This is the reason you need to get a lien release and waiver forms at the conclusion of any project on your home.

What Constitutes a Valid Lien?

To be valid, the contractor must be in compliance with his state statutory requirements.

The contractor must give the investor notice of his intention to file the lien within a specified number of days of work completion and non-payment. The lien papers must provide specific details of the amount of debt and scope of work for which payment is pending, and the name and address of who authorized the work (investor or homeowner).

The lien needs to be filed with the county court or registrar of deeds within a number of days of work completion. The contractor must file a lawsuit to collect the debt, legally known as perfecting the lien, within a specified number of days of filing the lien. These conditions make the lien valid. Anyway, once filed, your best route is to seek legal advice, as the presence of a lien on your property clouds your title and you may not sell or refinance until you resolve the problem.

In most states, the law applicable to mechanic's lien on residential property affords more protection against involuntary sale than liens filed against commercial property. Even so, if the lien is valid, if the contractor's lawsuit is successful and no alternative to foreclosure is available, the contractor may force a sale of your home to collect the debt.

A good rule of thumb is to check with an attorney for your specific situation since mechanic's lien laws vary from state to state.

Workflow: Managing the Rehab Process

The initial procedure is to ensure that contractors pull the necessary building permits. You may verify which permits are required for which specific jobs through your local government. As the work starts, pay attention to:

Demolition, Cleanup, and Trash Removal

Contractors will demo and remove damaged structures like the walls, cabinets, floors, ceilings, bathrooms, and kitchen fixtures as necessary. There may also be outside demolition and removals of unwanted trees or limbs, fences, decks, bushes, and even stones.

Framing

This may include redesigning the walls, cabinets, restoring demolished structures, any new designs, floors, and other areas that may need a re-design. Contractors should also look at the integrity of the foundation and take corrective measures if necessary.

Electricals, Plumbing, and the Heating/Air System (HVAC)

These are areas that may mostly require a permit depending on the extent of replacement work involved. Ensure that, if permits are required, the necessary permits are pulled, followed by city inspections.

Insulation

This should be performed after HVAC, electricals, and plumbing have been done and have passed inspection. In some cases, another inspection may be necessary after the insulation is finished to ensure you have not covered any wiring, piping, and ductwork.

Scraping, Trimming, and Painting

Make sure contractors scrape loose paint before painting. Some contractors paint over the loose paint, and that starts cracking within a short time, sometimes even before move-in. Trims should also be done before painting, unless the trims are pre-painted.

Exterior

Outside of the house should be done according to the scope of work and contract.

Walk-Through Inspection and Final Payment

No matter how meticulous you are, or how much you trust your contractor, there will be a few things that go unnoticed. To provide for this, it's necessary to conduct a second walk-through of the property, after all initial inspections have been completed.

Make sure the contractor delivered everything listed on the contracts. Also, don't forget that final inspections need to be done to finalize your building permits, and obtain Certificate of Occupancy according to individual state and city requirements. When you find that the project has been finished according to your specifications, draw up the final lien and the necessary waivers to be signed by the contractor. Some contractors will prepare their lien release, sign it, and give it to you. Once you are satisfied the contractor has complied with all the contractual obligations for the project, promptly deliver the final payment. That ends the contract and the job.

Handling a Rude and a Bully-Type Contractor

However, sometimes the contract doesn't go well, and you and the contractor may part ways prematurely. Sometimes, the contractor is not well-behaved and communication doesn't go well. It's a good idea to be prepared to handle a contractor who openly does not play by the rules all parties have agreed to in the contract. You need to identify his untoward actions as soon as they begin to manifest so you can handle them promptly before you become overwhelmed.

A friend of mine in the industry once remarked that some contractors do have one more trick in the bag they can always pull on you. Sometimes, you think you have seen it all, then comes a behavior right out of the left field. As an investor, you have to be quick, and if you get behavior you are unable to handle, then mention to a member of your team or seek professional help through consultancy or mentoring.

Matt started as a pleasant and knowledgeable contractor. He aced the interviewing process. His job references were excellent. He spoke well. Awarding him a contract was a no brainer. I paid him the initial payment, and he started the demolition work well. The first week was great. He and his crew aced the demo. I was super impressed. He then told me he needed more draw money than we had agreed on so he could do the HVAC, the electricals, and the plumbing. He mentioned that it was a blessing to have these professionals all lined up. He said that rarely happens, so we are on to fantastic start. That meant we could finish way before our schedule.

In my excitement, I agreed and paid him more than the contractually agreed-upon first draw. We were now outside our planned draw schedule. Big mistake. This was the beginning of our problems. I had been outwitted and was now trying to figure out how to catch up. Matt started missing work. He came up with excuses. His uncle had cancer. His wife was sick. His truck was in the shop. He had been robbed. The subcontractor professionals he had lined up were still finishing the projects. This was insane. He was driving me nuts.

When he finally showed up after two weeks, he came in as a fighter. He knew I had caught him with his hands in the cookie jar, and instead of being contrite and coming in to apologize, he blamed me for not being sympathetic to his situations, for pushing him to the edge on acts of nature and of other subcontractors—actions over which he had no control.

I explained that things happen and I am under the gun from the financiers. I mentioned that I had paid him more than the agreed-upon draw. I also said there was no need to keep talking. It was supposed to be action time on his part. I had more than fulfilled my end of the bargain. It was now his turn to perform.

He continued to blame me, saying he's never worked with such an uncaring investor, that I was a rookie and needed to learn to relate with people—specifically, with contractors. Matt was in his sixties so he talks authoritatively, almost addressing me as his child:

"My friend, my son, listen to me. At my age, I have been at the block, around the block, out of the block. You name it, I have seen it all. There is nothing you are going to tell me that I have not seen. I know what I am talking about. I know when to push, when to relax, and when to let go. Seems like you have no clue what you are talking about. You are getting mad because I am two weeks behind."

"I am going to tell you one thing. Contractors have so much power and if you push them carelessly, you will be a marked person in the community and no contractor will want to work for you. I am a trusted fella in our group, and if I disown you, you will be disowned by many."

"My brother, take it easy, your work will be done and I will finish before schedule. I am a professional, so just give me time. You have paid me the money. I know what to do. I have many projects under my belt. Yours is a small one," he said as he looked me derisively as a father lecturing a son.

I played cool, then asked him if he had finished. It was my turn. I referred to him as a brother.

"My brother, I hear you. I have heard you. I'm not as new in this business as you think. But, that's not the point here. Let's stick to the work that brought us together. You came to me to pay you more, and I did. Therefore, I believe, in my humble opinion that I did what you asked for. I have never done that with any contractors, so please know that I have treated you with the utmost respect you deserve. Please know that the money I am using for this house is not mine, but from a bank and other lenders. Please understand that I am under the gun. To cut a long story short, let's move on. Please let's write out another schedule of how you will proceed until your work lines up with the draw I have paid you?"

He looked at me derisively again and asked me to trust him. He would not give me a straight answer of when he wanted to proceed. He said it depends on the subcontractor's availability. I mentioned that it's absurd that when he came for the draw, he said emphatically that his team was all lined up and ready to start work.

They were ready to start like yesterday. After getting the money, he now could not tell me their schedule. He asked me to give him another week and he will bring everyone in. I wanted to put everything in writing. He told me to have some trust in my heart. Basically, he refused to provide a revised work schedule.

In the past, I had dealt with unapologetic contractors. We fought over many things, but I always stuck to the terms of the contract. This was a new territory for me. The contractor had pocketed much more than he was capable of delivering according to our draw schedule in the contract. My confusion in paying him more was based on his age, the way he interviewed, and the way he worked in the first week.

I was in a quandary. What else could I do, I wondered? Everyone I spoke to told me to give him time, that two weeks behind is not a make or break. I became patient and gave him time. As promised, he came back with an electrician at the end of the week. They worked for two days then Matt came back to me for more money. His claim was that the electrician wanted to do more wiring than we had budgeted for. I said no, and the electrician didn't show up again. This time, I was on a warpath. I was ready to call off the contract. I mentioned his breach of contract by not showing up, constantly pushing for more money, being hostile when corrected, and using demeaning and unprofessional language. He also vocalized a threat that I would be blacklisted among contractors and no one would want to work with me. In my book, all these are red flags that a contract should be terminated.

His behavior fit that of a bully, trying to use his age as an advantage and vocalizing threats like if I continue pushing him, he will quit, then sue me or file a mechanic's lien. In my dealings with him, I always remembered that the job is mine, so I was careful not to get in the gutter with him.

Although he didn't threaten to sue me or to quit, his saying that I would be blacklisted in the community when he was the one who was wrong was enough to gauge how our future work would proceed. It was time to cut him loose. Despite that, I gave him some time as others in my team had advised. Matt was one of a kind. He sometimes raised his voice. Sometimes he was extremely pleasant. He had many tricks. One day, he decided to bring another contractor to put up the outside railings at his expense. This was not part of his job, but he did such little things for free, when we were still stuck in the bigger planned projects like the HVAC, electricals, and plumbing. Every time I pressed him to talk about these big projects, he became loud and told me how unappreciative I was. He said I should see that he can do things, above and beyond, and not ask for payment, and so I should trust him even with the big things.

At one point he became so loud, I feared for my life. I stood my ground that he needed to bring another electrician who could do the job, or we would break the contract. He said he didn't want to be dealing with ignorant people who, because they have money, keep looking down upon the contractors. I saw his demeanor and I realized that as the project owner, I had to be the adult in the room and calm things down. I knew it was my responsibility to disarm him with a charm offensive. I gave him a bottle of water, and told him to take a deep breath. I advised that we re-group the next day early in the morning. He accepted and left the premises.

In my years in the business, I had learned to return aggression of any kind with calmness, meet hostility with composure, and focus the discussion back to the written contract. When Matt came back the next day, I refocused him to the contract. I told him what my obligations were under the contract, and what his was. I reminded him that I had over fulfilled mine. He still needed to fulfill his. He was calm. He came up with suggestions. He contacted another electrician right there. This electrician came the next day, and the electrical work started and proceeded to completion without a hitch. Matt still owed me for work on HVAC and plumbing.

After the electricals, he scheduled for the HVAC and plumbing. In working with Matt, I established my long-held belief that when working with bullies, do not ignore any contractor's strange behaviors. If I had not been persistent on him to return to work in the first week, he would have moved to another job as I later learned. His bullying tactics were meant to test the waters to see what he could get away with. He was testing my mettle. He found me grounded and firm. He had nowhere to run. If he was hot, I cooled the temperatures. When he cooled down, I refocused our conversation back to work. I didn't react to his abusive language. I took extreme care to make sure he didn't distract me.

Matt and his crew worked extremely well from this point on. I had two projects running concurrently. While I was battling Matt, I took my eyes off the ball on another small rehab project. Contractor Eric was another one. His job was to refurbish the whole kitchen in the property he worked at, carpet one bedroom, then paint the whole house. By the time I hired Eric, I had done so many projects for myself and others that I had learned when to be soft and when to be rough. Since I have a gentle personality and speak softly, many contractors also see that as a weakness, believing they can walk all over me. Not really!

Eric was a well-spoken gentleman, but had some audacities I could barely explain. He always asked in a conversational manner when I would be back at the site next. I never read anything sinister in this. What bothered me was the frequency with which he asked. I started smelling a rat. I had learned earlier in my career not to ignore strange contractor behavior, no matter how small. Some of these could be the contractor's way of trying to sweep poor workmanship and other quality issues under the rug. In this case, Eric had known when I would not be there to literally sweep poor quality work under a rug. One of his jobs here was to remove carpet from one bedroom, see if the supporting floor was intact, then put new carpet. If the underneath floor was weak, he needed to strengthen that first—may be cut out the weak parts, strengthen the whole floor, then carpet.

What surprised me when I came back to the property was how quick the work had been accomplished. Eric and his crew had painted the room where the carpet needed to be put and put the new carpet, all within the span of a couple of hours. I didn't believe it, so I returned to the worksite in the evening after they had left, and found the unthinkable. They hadn't scraped the walls before painting. They had put new carpet on top of rotten floor sections. I was mad since I had always treated Eric as a real professional and a straight shooter. Boy, was I disappointed! It was getting hard to find a contractor I could trust one hundred percent, or even ninety for that matter!

I started thinking that perhaps I should be realistic and lower my expectations of contractors to seventy-five percent. That way if I can get eighty to ninety percent, then that would be excellent. Perhaps this would be more realistic than thinking perfection. Lowering expectations this way would probably reduce my stress levels.

Usually, when I did my inspections when a contractor was not in, I always left marks on the place to be worked on. In the case of this floor job, I pulled all the carpet out. So the contractors knew I was there and that work needed to be done right. Whenever I do this, many contractors would know they have been busted and would do the necessary. A few others would be rude by saying I didn't pay them enough. Others would correct the mess and blame it on the subordinates. I know the game, so as long as they fix it, I will tell them to let me inspect in different stages as they progress.

In my dealings with Matt and Eric, and others, people with different personalities, I learned that no matter who I was working with, bully or no bully, my responsibility was to be firm but reasonable. I learned not to give in. Sometimes I took strategic retreats like I did with Matt, but stayed on task. The key is not to let the bullying words of contractors threaten or sway you. Your money is on the line. You own the project. You always have to remember that you own the job, and the risk remains with you at all times.

Other Lessons

Along the way, I learned other important lessons. Don't trust contractor's words without verifying. As

President Ronald Reagan once said, "Trust, but verify!" This is a necessary concept in rehabbing. I always reminded the contractors that after they are long gone, I still have the houses and tenants to deal with. The repairs will be on me. Tenant complaints will be on me. I make the contractors in my projects and those I manage aware that I am the final authority on matters of rehab.

Also, my advice to new investors is to find a way to talk with subcontractors when the contractor isn't present. Ask if they're getting paid and if there are other problems on the job. You know if they are not being paid well, then they will mess up the jobs. Even worse, they could eventually file a mechanic's lien against your property. If a contractor's behavior starts changing without no proper cause, then at an appropriate time, pull him aside and ask why he has been behaving differently lately. Ask if there's anything you can do to correct things for him. Maybe it's a simple misunderstanding or miscommunication.

The best way to avoid problems is to thoroughly vet prospective contractors. Narrow your search to companies that have positive reviews on a trusted online site, are appropriately licensed and insured, and work from detailed contracts. Take time to contact references.

In all cases, try to document all your interactions, so you have enough evidence in case you find yourself in a troubling situation with someone you've hired. Document communications and interactions using emails and texts as much as possible. In case of an actual threat, you may consider calling 9-1-1 if you cannot deescalate the situation.

If dealing with a licensed contractor, file a complaint with the licensing agency of the state. The state may mediate the problem. You can also file a complaint with the attorney general of your state, and they may try to mediate the situation. Also, you may consider terminating the contract for breach, in case there is a clearly identifiable infraction. Make sure you comply with your state laws so you don't invite a lawsuit.

Throughout this chapter, you have learned the elements of a good contract and how to navigate through the various hills and valleys when handling different contractors from the beginning of the project to the end.

Chapter 10:

Contract Completion and Release

The work has been going on for the last six months. The end has come. The contractor has come to your office saying he is done. "My friend, the job for which you hired me for is finished," general contractor Marty said as he walked into investor John's office. I was with John helping out on this project and others.

"I'm ready to move to the next project you have for me," he joked. "I think we have worked well, save for a few minor hitches here and there, but that is normal on a job of this magnitude. Don't you agree?"

He knows I work hand-in-hand with John, so he felt free to say whatever he wanted to say.

John didn't know what to say even as Marty talked and looked him directly in the eye. I believe Marty was trying to gauge John's body language, as they had truly gone through a lot. In fact, when John approached me for help, kind of a mentoring/consultancy situation, he felt defeated. Marty had been driving him nuts, constantly involving additional work that increased the original contract amounts. He stretched projects to appear like the work was harder and taking longer than planned. His game plan was always to figure out ways to keep asking for money.

When I came in, about four months into the project, I found when John and Marty were barely on talking terms. I started by listening to John's side of the story, and reviewing the contract and financials.

Then I went to Marty, introduced myself as one of John's key financiers, and that I would from now on be working hand in hand with John until the project was complete. Marty requested something in writing, preferably as an addendum to the contract so he would be clear on who to directly work with and report to. I mentioned that we will put an addendum to the contract and all the parties will append their signatures.

John and I made changes to the contract immediately, and presented to Marty. He agreed with the amendment to the contract, and off we started. I listened to John's side of things, why he kept doing additional unauthorized stuff that not only busted the budget but also prolonged the time to completion.

Marty pointed to the work that increased the budget. They needed to vent the bathroom and kitchen exhausts to the roof. This was a new city regulation, and inspectors would not pass the plumbing job unless the upgrades were done. He pointed out that they overlooked this at the contract stage, and when he brought the issues to John, he would not listen to him.

I didn't tell Marty my views, but went and told John that the contractor was indeed right. His approach of looking for additional funding may have been loud and wrong, but the request was justified. He should have explained better and convinced John of the necessity of the upgrades.

Since I had realized that communication was the key reason these two didn't get along, I started with sharpening the areas of communication between these two parties. I spoke to Marty and John separately. We agreed that communication would go through me from that point on. Both agreed, and we also re-iterated that official interaction from this point on will be in writing. For example, if the contractor needed money, then he had to do that in writing—text or email. If he wanted inspection so he could get draws, he needed to do that in writing also. If we wanted to correct something, we did that by text or email. We also agreed that inspections will be done jointly by John and me.

From my observation, the two people had great chemistry. They both seemed honest. What was missing was a streamlined way of communication. This is what I brought in, and the work proceeded well. Had I not been invited to help, these two personalities would have been at each other's throats. This was an example of how proper communication through properly laid down channels can contribute to smooth workings in a project.

We jointly worked on this project and saw it to completion. On the day Marty came to announce the completion of the project in John's office, all of us could hold our heads high that we all cooperated and we had something to celebrate. As Marty declared the end of the project, and joked that he wanted to know what the next step for him was, the happiness on his face and ours was unmistakable! John was happy, but not completely enthused! He believed the job had ended well, but not as smoothly as he had anticipated.

I kept trying to convince him that Marty was a contractor worth keeping. He's as good a contractor as they come. "You are not going to get one hundred percent perfection," I emphasized. "Not with contractors. Not with anybody!"

John seemed to agree, but with a heavy heart still. I told Marty we will discuss the next projects, if any, later. We needed to close out this job. I explained that on this project, the next step was to double-check that all the written and unwritten contractual obligations have been complied with and finished. We needed to make sure all the bills he was responsible for, including for subcontractors, had been paid. Any other disputes, if any, were to be settled according to the contract. Thank God, John had invited me on time. There were no pending issues to be sorted out. We set the final walkthrough for the following day.

By 9:00 a.m. on the dot, the three of us were ready to move on with the walkthrough. John brought his final checklist, which he and I had previewed the previous evening. At the end of the walkthrough, we jointly put together a final punch list. This covered securing the microwave well and making sure it was positioned upright. All the loose sockets needed to be secured and second coats of paint applied to the areas we pointed out. Upstairs bathroom vent was making strange noises, so it should be taken down and re-installed. More caulking was necessary in several sections we pointed out. Marty needed to tighten the locks in various doors and repair the loose steps at the basement staircase. The rest was minor stuff. Marty said he needed two or three days for the repairs, then we would do another walkthrough, and move on from there.

He finished the repairs in two days. Then we did another walkthrough on the third day. Everything was copacetic. With this walkthrough, all the three of us agreed that work had been completed according to the contract. We shook hands, and I congratulated Marty and John for persevering through the whole building process and coming out with a great job. I mentioned to John that this is the one contractor he needed to keep in his fold. I reminded John that rehabs can be a pain for all parties, and he did a good thing by bringing me on board when he did. I told Marty to understand where investors are coming from. They have a tight budget from which to accomplish the tasks. They are also accountable to the city, financiers, and others who may have interest in the projects.

With everything in order and John satisfied that the contractor had met expectations, it was now time for Marty to give John a release document from himself and release waivers from his subcontractors and suppliers. John said he already had the subcontractor and supplier waivers on file. Marty did not have a release document drawn out, so we printed one from the internet. He filled out and signed. The release and waiver documents stated that he, the contractor, and his employees or subcontractors or suppliers have no other claims on the property as they have been paid in full per the contract. With all that squared away, John paid Marty his final dues by check as they had agreed. That closed the project. John ordered pizza for all the workers and we celebrated this great run in style.

By many accounts, this was a project that ended well. However, sometimes it's not this straightforward. In many cases, you will find that you and the contractor don't agree on some things. You always attempt to solve the problems the best you can. If you can't agree and find that you are butting heads, then feel free to bring in one or more third parties to help with the communication process. In John and Marty's case, even though I came in as a third party to help with the communication, everything moved pretty smoothly. There were no other complications.

If the third-party intervention doesn't work out, then do a formal mediation. Make sure you are solving your disputes according to the contract conflict resolution clause. In some cases, you may find you are so far apart, that no matter what you do, you all seem to keep drifting apart. Under the circumstances, you can see if it is in your best interests to negotiate a way out of the contract without prolonging the pain.

Sometimes, the damage is so deep and the breach of the contract too severe that moving apart immediately is the best solution. You may have to fire the contractor. Firing may be the only way to minimize the damage or to stop the bleeding. Again, follow the contract and make sure you are in compliance with your local and national laws. You may even consult with an attorney if you feel the contractor may try to file a lien or sue you outright for breach of contract. Be careful, but do what you have to do.

It's Okay to Fire Your Contractor for Cause

Firing a contractor mid-stream can be destabilizing for both the contractor and the investor, but if the contractor is not qualified to physically perform the job, cannot be relied on, and does not communicate well, then it may be time to part ways. When you make a mistake and hire a contractor who doesn't work well and is unreliable, it's in your best interest to part ways sooner rather than later, so long as the contract you both executed included a termination clause. As some people may say, it's okay to be wrong, but it's not okay to stay wrong!

The termination clause in the contract should state that the contract may be terminated without cause if certain criteria are not met or accomplished within the stipulated timeframe. For example, a contract could state that if no contractor shows up in a job within four consecutive days, without cause, at any period after signing, then the investor will have the right of termination. Many contractors have the habit of booking more than one job just to get initial deposits and basically hold different investor jobs hostage so another contractor cannot be hired, then they show up for each job as they decide. Investors should look out for such games.

A good contract should also contain a no-lien clause so subcontractors may not put a mechanic's lien on the property should the main contractor not pay them. Also, the contract could include a mediation clause so the parties can arbitrate or use the court process as necessary. Both parties should clearly define the process and what it takes to trigger a warning, arbitration, a termination, or a court process. All these can be avoided at the hiring stage. Although there are no guarantees in life, when a clearly defined hiring process is followed, then there should be less disturbance once the job begins. Also, after hiring, the investor must stick to the contractual agreement for the whole rehab process to flow well.

Many investors fear speaking up when problems first arise because they don't want to upset the contractor. That is wrong. You should communicate as soon as you notice a deviation from the contract. Whether that is lateness, drinking on the job, or stealing your materials. Handling a rehab is such a huge undertaking that fear should not be part of the equation. Speak out, just know how to communicate so the contractor doesn't feel you are looking down upon him or his crew.

Don't Hire Someone You Cannot Fire

To avoid complications, don't hire someone you cannot fire. Period! That includes a relative, a friend, a friend of a friend, an ex, or a bully. Also, do not fear to correct a situation as it crops up for fear of losing your investment. From my experience, problems have a way of multiplying when not taken care of when detected. In many cases, such problems will increase and get worse.

Structure Your Business Well to Eliminate Fear

Some business owners have this fear of being sued if they correct bad business conduct of their employees. As an investor, please protect yourself and your investment on all fronts by having a watertight contract and by hiring the right people. You can also consider buying general liability or umbrella insurance to cover yourself in case you hire a contractor who does not have adequate coverage. The contract should include clearly defined goals, yardsticks, and expectations so all your bases are covered, so you have nothing to fear when conducting your business.

Work and daily progress should be documented. The daily progress should include materials bought, materials to be bought, and daily goals. Incidences of non-achievement of daily goals should be documented and corrections made. The contractor should make and keep the records in a place that is readily accessible during or after work. You should feel free to discuss with the contractor any concerns you may detect at any time.

The goal of a contract is not only to make sure work proceeds according to plan, but also for both parties to adhere to all their responsibilities and obligations. When all the parties stick to the terms of the contract, the working atmosphere is peaceful and businesslike. When either one or both of the parties decide to dishonor their side of the bargain, then the result could be disagreements, mistrust, and even premature termination of the relationship. The investor should keep his end of the bargain by learning the nuts and bolts of the business and sticking to the contract as much as humanly possible.

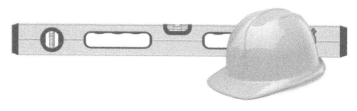

Chapter 11:

Obstacles, Obstacles, and More Obstacles

Starting any business can be challenging. Starting a real estate business can be even more challenging, due to initial high dollar investments needed and the many moving parts involved. The moving parts are the nuts and bolts that go into buying, rehabbing, renting, or selling a property. These include financing, city and other government regulations including permits and taxes, legal and accounting issues, contractors and contracting, managing the rehabbing process, professionals like the realtors and brokers, stagers and cleaners, and relationships with supply and home building companies. Proper planning, budgeting, and record-keeping, combined with clear communication among the parties for each moving part ensure that the parts connect seamlessly into the whole.

As this book has covered in the earlier sections, for each moving part to fall in place, meticulous planning and execution should be followed from the purchase of the property and materials to rehabbing, renting, or selling. The planning starts from the investor's familiarity with the legal and building landscape, as there are steps that need to be done in a sequential manner so as not to flout the city and government regulations. For example, permits have to be pulled before commencing work in certain sections of the building according to city guidelines. Inspections then follow before finishing those sections. I have seen situations where inexperienced contractors seal some permitted parts before inspections, then they are forced to open these sections up when inspectors come, and re-seal after passing inspections. This type of inexperience will cost an investor in terms of time and resources. Contractors don't like being forced to open up the walls even though it's their fault. These contractors tend to figure out how to burden the investor, either by increasing the cost of performing some parts or trying to negotiate more time to finish the project.

Such delay and headache can be minimized if the investor understands the work, negotiates an inclusive contract, supervises, and manages the project well. The main obstacle in implementing the rehab process is overcoming a lack of knowledge by the investor. This lack of knowledge can be costly and takes time to master. Investors who may experience this knowledge obstacle are mostly the ones who are new in the business or who started the business without proper grounding. Sometimes, the toughest people to change may be those who have learned things the wrong way. As some investors with many years of experience say, this may be a situation of trying to teach an old dog new tricks. As one of my professors put it: good luck with trying to change the ways of a grown man. This professor's solution, which I partially disagree with, is to always hire young and new people, teach them the right way, and you will reap the benefits for a long time.

The professor and I are in agreement that, if you get a clean slate, teach them the right way, and instill good habits into them, they will mostly grow and perform that way. My disagreement with him is on two fronts. First is that I never want to ignore the power of experience, which is crucial in the building industry. Second is that, in this day and age, people are mobile, and just as companies may no longer have long-lasting loyalty to their workers, the same goes for the workers about their company. The young people, more than most, tend to have less loyalty. To my arguments, the professor would say the best way is to have a built-in employee retention yardstick, like bonus pay for beating pre-set deadlines.

In the real estate world, a way to hold your exceptional performers captive is to either partner with them in some new projects or promise them a bonus if their exceptional work leads to higher than expected sales. Other investors give their contractors a percentage of sales profit beyond a certain level. All these can be negotiated in the contract. Care has to be taken to offer these incentives, because some contractors may compromise quality to finish faster or even do cosmetics that provide more appealing structures to buyers and compromise quality. Poorly rehabbed houses like these may start falling apart within a short period after purchase. Again, all these incentives can work well when an investor invests in his education and progress as circumstances change in the industry.

In case you have been in the business for a long time and are still stuck in your old ways of doing business, or in case you are self-taught and constantly on a slippery and sliding slope, then you may need to take a strategic break and learn. Better yet, you can hire a reliable consultant to show you the new ropes before going back solo again. You have to decide what works for you. All you need to know is that a lack of proper education in the ever-changing industry environment can be a big obstacle to your progress in the business. You have to be able to evaluate where you are and be willing to take corrective measures to straighten out your situation.

Sometimes, outsiders may help you better evaluate your situation and your business progress more than you can, given your built-in biases or unwillingness or inability to change. It is best to know that you can be your enemy or obstacle to building or re-building a thriving business. If you can't do it alone, hire someone to help you.

Resistance to acquiring new knowledge can be a big obstacle to your business progress. The other obstacle is the issue of comprehending this new knowledge. While you may be willing to learn and make financial and time allocations for education, the question could be whether you can understand the new ways of doing things. Some of these new ways may involve outsourcing some of your services, like accounting and legal work and even buying and selling properties.

This would allow you to concentrate on doing what you believe you do best. If that is working with contractors, then that should be it. If it's finding resources to pay the contractors, then that's what you should concentrate on. Sometimes, it is not easy for one individual to understand it all. That may be a good reason to hire others or outsource what you don't enjoy or are not good at. If you are not good at hiring and managing contractors, then have a consultant or mentor work on that, or work with you on that. You will find that you are enjoying your work, and there may be less friction if you specialize in doing what you like and are good at.

As mentioned earlier, another key moving part in the real estate business is the issue of hiring contractors and managing the work process. In reality, hiring contractors and managing the workflow is the backbone of the rehabbing business. This is the thread that runs through all the building processes. You need a contractor to consult with on the buying of property and materials, building the house, and even the selling analysis. Contracting is a huge part of real estate investing. You need to get it right. Therefore, if human resource management is not your forte, then outsource the hiring and managing contractor relations. Contractors come in many fashions, and so outsourcing the management part might be one way to handle them if you can't talk well or you lack patience. For communication to flow well, you need to set ground rules that all parties agree to in a contract and follow without deviation, unless absolutely necessary.

If the contractor understands clearly that you are running your business as a business, they will treat your business and interactions with you in a business-like manner. If you become casual and accept deviations without following the letter and spirit of the contract, then they will treat you in a casual manner and cut corners as much as they can. The rule to beating contractor mischief, if detected, is to be business-like and firm. You need to up your human resource management capabilities through more business education; otherwise, hire consultants until you know what you are doing!

Without much thought and knowledge of how best to proceed in the hiring and management of the contracting process, you are mostly going to encounter obstacles. Even seasoned investors do encounter challenges with hiring contractors. As a new investor or one still struggling with the hiring process, you need to be aware of the potential obstacles and pitfalls that may be in your path to hiring the right contractors. While some investors use consultants, others get mentors. Other stubborn ones still go it alone with disastrous consequences.

Starting a new business without education, experience, or help can be fraught with challenges, and you need to be aware of the obstacles you may encounter. Some of these obstacles may include lack of knowledge, lack of capital, being misused, self-doubt, confusion, not knowing when to hire or fire, and a wishy-washy management style. These challenges and obstacles can impact the contracting processes and interfere with all aspects of a rehabbing operation.

Many people who have run small businesses do say that finance and time are the greatest obstacles they face. You need time and resources to invest in your education. Education will improve your knowledge for analysis, inviting new ideas in your business, detecting current real estate trends, and incorporating new technology for property search, computerized business management, and record-keeping. Some business owners do not have time to attend seminars, webinars, or networking events. All these may be avenues for increasing knowledge so you run your business in a more streamlined manner. Some business owners spread themselves so thin, they never have time to absorb information that improves their operational efficiencies and increases their bottom line.

In the area of contracting, lack of proper knowledge of your project and inability to detect contractor tricks during and after the interview may lead to hiring unsuitable contractors and compromising the quality of your rehab. For example, a lack of knowledge on how to hire suitable contractors for your job could lead to defective criteria for choosing contractors. Improper criteria may result in improper shortlist of candidates. Improper shortlist of candidates may lead to untrustworthy contractors. The inability to understand real estate contracts may lead to loopholes that contractors may exploit. Inexperience in the industry may lead to mismanagement and workplace confusion. Lack of a clearly spelled out contract combined with mismanagement of the whole building process may increase conflicts, resulting in an imperfectly finished rehab.

Working capital is another area that creates obstacles in business. For projects to be done well and contractors paid properly and on time, an investor needs to have ready access to capital. This could be from personal resources, partnerships, equity infusion, and lines of credit. You can learn how to source funds from networking, attending local real estate association meetings, and visiting various lending institutions like banks, hard and private money lenders, family, and friends. Your goal is to ensure that your project does not stall due to inadequate funding. When your project stalls, you will be faced with a lack of continuous operations, and some of your reliable contractors may pick up other jobs and may not come back to you.

There have also been instances where an investor runs out of money, and the contractor took him to court for breach of contract. The contractor's argument was that the business owner was negligent for not foreseeing and including a language in the contract that he may not have adequate funding throughout the project, and that this situation may interrupt operations including payments for labor. The contractor argued that had he known this lack, he would have budgeted his income accordingly and not involved himself in other personal investments. He also argued that there was a possibility he would not have accepted the job in the first place. Alternatively, he would have arranged for some jobs so he was fully occupied in times of work interruptions with this investor. As such, he was suing for full payment of the contracted amount including penalties and damages for loss of income and general interruption in his personal life and investments.

The investor had to borrow money to settle with the contractor outside of court. Otherwise, he would have paid this contractor huge penalties for loss of income and his other personal lost investments.

Sometimes, as an investor, if it feels you might run out of funding, you can reach out to consultants you may have worked with and they can direct you to places where some of their clients get construction money. Consultancy or mentoring can be helpful in many respects. As a businessperson, don't be fearful or proud to ask for help. Get help whichever way you can so your project doesn't stall. You start losing respect among contractors, and your name may get blacklisted in a community as someone who makes agreements with workers and doesn't pay them. This is a field where if you give respect, you get respect! Sometimes, as an investor, you may give respect but may not necessarily get it! You either ignore or move on! You don't fight unless you absolutely have to. There is too much at stake to be bogged down by unnecessary arguments. You build a reputation in the industry, and people who come to work with you learn to treat you with the utmost respect you deserve. Once you build a name in the business, hiring the right contractors and managing them becomes relatively easy. A good name will also get you referrals, and you will give referrals. That's the beauty of becoming a known entity in the business, whether locally or nationally if you get to that point.

Investor Obstacles as Told by Contractors

As a research student in graduate school, I always enjoyed digging deeper into issues and situations. Even as I zeroed in on the above obstacles that a new investor or even a seasoned investor may encounter, I knew I needed to hear the contractors' take on the obstacles they believe investors face. My research revealed varied and interesting answers, some clearly unprintable. Will leave those out!

What I learned from the get-go is that there is a deep love-hate relationship between investors and contractors, and many contractors have little sympathy for investors, even though the contractors believe they sometimes put the investors in no win situations. While both sides admit they need each other, they think both sides play games, and many contractors say they have no time for the investor's shenanigans. The contractors believe that investors do underpay them in many cases. They also believe the investors do not respect them, even though many of them cannot build the houses without the contractors.

Despite all this back and forth, I convinced the contractors I spoke to that both sides need each other, and so it's important that both strive to understand each other. I wanted to concentrate my research on the general areas that may create obstacles to implementing the framework for hiring reputable contractors who will get the projects done on time. An assortment of various contractor views is outlined below.

LeRoy, a general contractor in Baltimore, Maryland for over twenty years, said an investor, first and foremost, needs to know how to accurately define what he wants the contractor to do for him. He needs to state this clearly and unambiguously. He can only learn this from talking to many contractors and other real estate professionals, like realtors and mortgage brokers. Leroy said there are many times when an investor gives him a job, only to change his mind midway, and then change his mind again. He said this is frustrating and gets people back to the drawing board. In many cases, such changes of mind call for additional money, which investors rarely pay willingly. This creates confusion and unnecessary friction, resulting in a poorly finished project. He says, investors need to know their projects intimately before approaching a contractor and signing a contract.

Justin, another contractor of many years, listening to my conversation with LeRoy, added that there are no two ways about it. If an investor wants to hire a good contractor who will finish his job on time, then intimate know-how of the project and the real estate environment around him will guide his hiring and contract management processes. Justin said he has worked with many new investors who have no clue what they want. All they have is some money. They don't even know how to price their work. They rarely know the full budget for their project. In many cases, they believe the limited money they have should do wonders, and if they can't get what they ignorantly believe they should get, they blame the contractors, and that creates lots of problems where there should be none.

Katherine, another contractor with some two years in the business, added her voice to the conversation. She said she works with many investors who should not be hiring her in the first place. She said people forget that real estate contracting is a job like any other. You are going to be working with someone for a long time, and you need to factor in your chemistry and his to see if you are a fit. If you are a strict personality, hire a strict personality. If you are casual, look for the same. That is because, for example, a strict personality will keep timetables and expect to be paid on time. An investor who has a casual personality may not have payment ready as planned. She said she worked for someone who was rarely ready with money when she wanted materials or draws. She had subordinates who needed to be paid, and this was a nightmare for her. She now looks for like-minded investors, and that makes the work flow a lot smoother.

These three contractors advised that investors should not take too long to move on the hiring process after identifying a good contractor. Good contractors don't stay long without jobs. An investor should go through all the items in his hiring checklist, but move faster. Procrastination, analysis paralysis, and fear of taking a new path different from the wrong way they may be used to can cost an investor big time as he may have to settle for lesser qualified contractors when the good ones he interviewed have moved on.

They advised that investors need to learn to be consistent in their actions. Many times, they work with investors who manage one project one way, then when you believe you are now in sync in the way you both operate, they change tack, and this creates lots of confusion. Payment scheduling is one such area. On one project, they may give a one third down payment. On another, they may argue for giving less than one third. They also said the way an investor hires should follow the same path so they don't overlook some areas because an interviewee is a sweet talker.

They said an investor should be firm and trust his intuition. An investor should develop this from research and experience, or mentoring. They said if an investor becomes doubtful during the process of hiring, he should ask someone to double-check to confirm his actions. He needs to avoid taking actions and then second-guessing himself, having excuses, or trying to justify his actions after the fact. He should have a mechanism of resolving his conflicting thoughts before hiring. A good investor should avoid having a negative mindset and associating with pessimistic people. If he is wrong, he should admit it, and see what compromise to make. He should accept that even though he has money, he doesn't have all the knowledge, and so it is important to learn to listen to contractors. At least, seek the counsel of those good contractors who have a proven record!

Mark, the youngest of the contractors in the group, only four months in the business, also sought to add his two cents. He said he understands that many contractors may have no time for technology and new machinery, but an investor needs to invest in them to cut down his costs. He advised that an investor need not be shy to use or fear new technology and new tools. Many older investors don't like change. He said investors should embrace change and see how to incorporate technological innovations in their business plans so they are not always operating from behind the curve. Once an investor is in tune with technology, he will be seeking contractors who possess such knowledge. This will cut down on his costs and time to finish a project.

The above obstacles highlight the roadblocks that prevent most people from solving their problems, resulting in them maintaining a status quo of constantly hiring the wrong contractors and frequently losing their investments.

By knowing the obstacles before you start in the business, you will get a balanced view to help you decide how to hire the right contractors who will finish your projects on time so you don't lose your investment. Being aware of these obstacles can help you decide if you want to solve the problem using my methods outlined in this book by yourself or if you want to do it with me. If you choose to use this book to solve your problem by yourself, be aware that most successful people don't do it alone when they start a business, or when they find they are in a jam.

Chapter 12:

So, What Have You Learned?

F rom my experience and from talking to many investors whose businesses are struggling, you get the sense that many investors confront various issues in the rehabbing process. Their one overriding challenge is the struggle to hire the right contractors for their intended jobs.

Shortlisting, interviewing, and hiring the right contractors is a major dilemma many investors never seem to shake off. Not solving the problem comes at a cost though, and may mean the success or failure of an investor. These costs involve hiring contractors you may not trust, a disorganized work environment because you don't have the right procedures in place, and losing your investments. This book goes through the processes that you can master to help solve this nagging problem of hiring the right contractors for specific jobs. You also got to see the opportunities that will come your way after solving this problem.

At the beginning of the book, I introduce you to investor John Jacobs, who is desperate and seeks to be enlightened on how to find good contractors who'll get his jobs done on time so he doesn't lose his investments. His problem is that he is always hiring the wrong contractors who lie often. In Chapter One, you were able to understand and identify with John's stress of hiring the wrong contractors and the pain of a constant loss of his investments. I let you know that there is a process for you to realize your dream of hiring great contractors that get your real estate projects done on time. The fulfillment of this dream brings ease and confidence to the process of hiring suitable contractors who will finish your projects on time so you do not lose your investments. In Chapter Two, I shared my story of struggles with contractors. I've been in the trenches and fully understand where you are coming from, and you can gain from my practical experiences.

Chapter Three outlined the road map that takes you from a place of insecurity because of constantly hiring underperforming contractors to a level of confidence as you master the methodology of finding and hiring reputable contractors. One of the first steps in hiring the right contractor for your job comes from acquiring a good knowledge of the real estate landscape in which you operate. In Chapter Four, I mentioned the types of styles and structures to learn that set you on the road to understanding your environment. You will combine understanding your environment with developing an intimate knowledge of your project, as Chapter Five outlines. A good knowledge of the job you want to get done is one of the key steps toward the type of contractor you would like to hire.

Chapter Six explained how knowledge of the local real estate environment and your specific project leads to a clear and defined way of zeroing in on a shortlist of quality and suitable contractors to invite for interviews. With quality candidates shortlisted, Chapter Seven showed you the various ways to prepare so you can approach the interviewing process from a position of knowledge and confidence. Hiring a suitable contractor is one of the cornerstones of a good rehab project. Hiring a suitable contractor involves getting good referrals and references, digging deep into the contractor's background, relevant work experience, and history. You also need to check the contractor's licenses, bonding, ratings, and any complaints.

Chapter Eight explained the best ways to evaluate the interviewees, eliminate those who don't meet your requirements, then hire the most suitable ones. With your already defined criteria as your checklist, you should make sure you meticulously follow your rules. If none of the candidates meet your criteria, do not hire. You should look for new candidates and start the hiring process from scratch. It's better to be right than to be quick! You need to remember that hiring mistakes can be costly, leading to poorly done projects and loss of resources.

After you hire the contractor who meets your requirements, you proceed to work on the contract and the work process. Chapter Nine explained the elements of a good and watertight contract. The contract outlines the business and contractual relationship between the contractor and the investor. The contract, if adhered to well, guides everyone's conduct, responsibilities, and expectations in such a way that gets the project done in a timely and businesslike manner.

Chapter Ten guided you through the processes involved with contract completion and lien release. As the contract is a legal document, you need to ensure that, not only is the construction finished, but the legal loopholes are also closed. In that sense, make sure the contractors and suppliers sign lien release and lien waiver forms so they do not come out later to argue that they were not paid in full and are, therefore, filing claims against you or your company. Throughout the book, the investor will learn many lessons that he can implement by himself. He will also realize that the road to that accomplishment is filled with obstacles.

Chapter Eleven highlighted and explained the obstacles that can prevent even the most experienced and seasoned investors from fully implementing the good practices of hiring contractors mentioned in this book. These obstacles cover lack of adequate knowledge in various areas, resistance to technological changes, refusing to outsource when someone becomes overwhelmed, lack of adequate working capital, and elements of procrastination and analysis paralysis. For you to achieve your dream come true, you need to burst through these obstacles to see the value in solving your main problem of mastering the art of hiring the right contractors. Solving this problem gives you the opportunity to run your business from a new level of renewed optimism. This optimism comes from knowing that your business can start making money, for a change.

John's Growth

Of the people I have worked with or mentored, John has recorded the fastest and most consistent growth. Even though his learning was full of ups and downs, he grew tremendously in the business. When I met him in 2018, he had been in the business for many years, but was still making rookie mistakes. He didn't like approaching other investors for fear he would not know what to say. He didn't have a system for hiring and retaining great contractors. He lost money in his investments.

One year after starting to work with me, he has worked within his budget and will be making money on this investment as his costs are much lower than the after-repair value (ARV) of the comps of similar properties in the area. He will make a profit on the sale.

He has now mastered the processes of hiring and retaining great contractors. He has also learned the contracting process, when to bring in a consultant, or when to fire a contractor, if situations demand. Above all, he has become a confident investor. The power of knowledge has given him a newly found authority that he wears humbly in his rehabbing career.

Conclusion

The results of my work with John have made him happy in his daily experiences with contractors. He now has a renewed sense of confidence in his business. Just as John has grown, you will also grow if we get a chance to work together. It is a lot of work, but the payoff is huge. Real estate is one of the industries where your investment could give you multiple sources of income, such as renting, selling, refinancing, and living in it. Rehabbing produces great income in the industry. Mastering how to hire and work with contractors can increase your success rate tremendously.

I hope that after reading this book you are now hopeful that the contracting process can be mastered and will lead to the hiring of the right contractors for the right jobs, at the right price, and for the right duration. By working with me, or a similar professional, you will acquire a repeatable, streamlined, practical, and well-thought-out methodology for hiring a quality contractor with a good track record. Hiring the right contractors who get real estate projects done on time will provide you with the necessary peace of mind for approaching the contracting process with ease and confidence. The book has equipped you with the necessary understanding to approach the contracting process by yourself, but you are also aware that interactions with contractors will be smoother and more successful with some professional help.

Acknowledgments

With grace and gratitude I wish to acknowledge the many people I have worked and interacted with in the real estate industry. Their stories and interactions form the bedrock of the foundational knowledge this book is built on. It is with this background, combined with the depth of my failure and revival in real estate, that I set out to write this book with the hope that I will meet struggling and new investors at their points of need, as I have been there, and can lead them to where they want to be.

To all the investors I have had the opportunity to work with, have mentored me, been mentored by me, or I have watched and listened to their joys and tribulations, let me thank you individually and also reveal that you are one of the many inspirations behind this book.

To all the contractors I have worked with, who have treated me shabbily, who have been good to me, and those I have watched work with others and listened to their joys and tribulations, let me similarly thank you individually and also reveal the same information that a good portion of my materials for this book came from such interactions.

The stories and corroborations in the book also came from investors and contractors I met at the real estate investment association meetings, meet ups, publications and from personal research. I hope a pleasant unintended consequence of this book is that contractors and investors get some idea of how they feel about each other!

To those who have written a book in the past, those who are writing now, and who will write in the future, let me be honest with you—this work is much harder than I ever imagined. The second guessing, the writer's block, and the feeling of personal inadequacy can prevent even the greatest writers out there from achieving their goals. However, surprise, surprise, please be warned that the joy of finishing a book and the feeling that you are now a published author, in case it's your first, overrides all the struggles you could go through!

Even more than that, the feeling that your idea may help another person, another human being in need of uplifting, encouragement, or just to get a breath of fresh air, makes the struggle worthwhile!

For many years, this desire to help others in any way I could, became an obsession, my mission, my struggle, and my dream. Brothers and sisters, it all starts with an idea. That idea can get a boost if you have a great team to assist you navigate through the maze of putting it all together. This is where the Author Incubator came in. I had been looking for a group that would help me incubate my idea and make it a reality. I found a home with this team, and I have never looked back. Therefore, a big, hearty thank you is in order here to begin with.

To Dr Angela Lauria, thank you for thinking and creating this organization that gives budding authors this transformational path to make a difference in their lives and the society. Many people go through various challenges in life and to provide them with a platform to develop their potential to be the difference and to make a difference on others is by itself the stroke of a genius.

To my editors, I cannot thank you enough for believing in me even when I had doubts about my capabilities. I have never experienced such overwhelming support, love, and guidance from an organization. I will forever be grateful to the following editors: Stacey Warner (Acquisition) for seeing the potential in me and my project, Mehrina Asif (Developmental) and Cory Hott (Managing), Todd Hunter (initial line editing, preliminary formatting, and the kind words of appreciation and encouragement).

Further, I wish to add a special thank you to Mehrina and Cory. I owe you guys big time! We worked so well together. Mehrina guided and constantly encouraged me in the process of organizing my thoughts into a coherent pattern. Cory made sure that I put together the necessary materials so he could do the line editing, formatting, and all the heavy lifting involved in finalizing a manuscript. His constant encouragement and appreciation meant a lot to my psychology.

To both of them, I must now reveal a secret I have held close to my heart for a long time. Many times through the whole writing process I had so much doubt of myself and my capabilities to finish the manuscript within the prescribed time. There were instances when I sought extension of time to finish. Mehrina and Cory came back and told me I had more than enough time. For a minute, I quietly thought these editors were out of their minds. As it turned out, I finished the manuscript a couple of days before the deadline. I was wrong. They were right! Their love for both the topic and my writings greatly inspired me to strive for excellence. To that, I say a special thank you to both of you for putting your feet down, and moving the process along to where we are: the end.

To all the others at the Author Incubator I interacted with, thank you for all your ever help and encouragement. A special mention goes to Ramses Rodriguez and Cheyenne Giesecke. To all fellow writers at the cohort, I am glad we walked this journey together. May you all be blessed in your efforts of seeking to make the difference you aspire to make!

Thank you to David Hancock and the Morgan James Publishing team for helping me bring this book to print.

To my very dear friend and business partner of many years, Sharisse Leacock, let me say I am eternally grateful for your support and confidence. You have been there through my uncountable ups and downs. Your encouragement and being there in many ways throughout the period of writing this book and before has been immeasurable. May the Lord continue to bless you abundantly in all your aspirations. You are a great human being!

To my granddaughter, Mya, your playful smiles and brightness always give me encouragement. You are such a bundle of joy and may the Lord continue to give you all the strength you need to grow into the strong person you are capable of being.

Finally, to those I have not mentioned here but feel I should have, please do not despair. This is my first book, and your turn will come many times in the future!

Thank You for Reading!

Writing this book opened my eyes and gave me an opportunity to think and reflect on the many moving parts in the real estate rehabbing and contracting sectors of the industry. Even though I have been in the business for many years, I still learned some things I have periodically overlooked.

As I approached people to test-market the idea of this book, the overwhelming response I got was one of excitement about the subject. Many identified with the topic in one way or another. A number of them said they couldn't wait to read the book. Others lamented saying they wished they had read such a book before buying and rehabbing their property. They said they would have avoided much heartache!

Therefore, to those who have read this book, I want to say a special thank you for your time. I hope you have enjoyed reading and gained useful knowledge that you will apply as necessary. If you still have problems with hiring the right contractors for your jobs, please send me an email, or contact me at the website below, and we can set up a time to talk.

As an appreciation for keeping in touch, and staying connected, I will send you a free gift that supports the book and highlights the framework for hiring the right contractors who will finish your project on time so you don't lose your investment.

Whatever the case, please take a moment and let me know how the book has helped you, or even if the book has stirred you to be better at working with contractors. I will enjoy reading your thoughts and the stories of your own experiences with contractors.

Like all authors, I appreciate getting feedback. To those who wish to write a review, or any form of feedback, which I welcome and greatly encourage as that is how we improve, please share your assessment on Amazon or any book review websites you choose. You may also communicate your feelings directly by email at contractorcure@gmail.com.

In addition, please remember to visit, follow, and like us at the social media sites related to this publication to keep abreast of new developments in contracting or other construction issues. Some of these sites are:

www.contractorcure.com

www.facebook.com/contractorcure
www.twitter.com/JOAMaurice1
www.linkedin.com/JOAMaurice

About the Author

A real estate entrepreneur for over fourteen years, J. O. A. Maurice researches, negotiates, buys, and manages a personal real estate portfolio in Baltimore, Maryland. He began in 2005 and experienced the highs and lows of the real estate boom that ended around 2008. Maurice partners with investors, big and small, to pool resources for the purchase and management of real estate assets instead of a massive reliance on debt financing.

A believer that knowledge and degrees lead to increased opportunities, Maurice has pursued educational advancement leading to an MS in Accounting from the State University of New York in Albany, a business undergraduate degree from Principia College, Illinois and many Semesters at a doctoral program in Business Administration at the University of Phoenix online.

His other professional experiences include working in the individual and corporate tax fields, business analysis and brokering, personal stock trading, financial advising, writing, counselling, and entrepreneurship.

Working in the rehabbing and renting field of the real estate industry combined with his other professional and educational experiences has prepared Maurice with capabilities to cater to the needs of investors in their dealings with contractors. These experiences have led to his work helping new and struggling investors learn to hire the right contractors for their jobs and help them manage their projects.

Lightning Source UK Ltd.
Milton Keynes UK
UKHW011830050221
378222UK00011B/320